Tom Adam

REVISION PLUS

OCR Gateway
GCSE Science B
Revision and Classroom Companion

Contents

Scientists carry out **investigations** and collect **evidence** in order to explain how and why things happen. Scientific knowledge and understanding can lead to the **development of new technologies** that have a huge impact on **society** and the **environment**.

Scientific evidence is often based on data collected through **observations** and **measurements.** To allow scientists to reach conclusions, evidence must be **repeatable, reproducible** and **valid.**

Models

Models are used to explain scientific ideas and the Universe around us. Models can be used to describe:

- a complex idea – like how heat moves through a metal
- a system – like the Earth's structure.

Models make systems or ideas easier to understand by including only the most important parts. They can be used to explain real-world observations or to make predictions. But, because models don't contain all the **variables**, they sometimes make incorrect predictions.

Models and scientific ideas may change as new observations are made and new **data** are collected. Data and observations may be collected from a series of experiments. For example, the accepted model of the structure of the atom has been modified as new evidence has been collected from many experiments.

Hypotheses

Scientific explanations are called **hypotheses** – these are used to explain observations. A hypothesis can be tested by planning experiments and collecting data and evidence. For example, if you pull a metal wire you may observe that it stretches. This can be explained by the scientific idea that the atoms in the metal are arranged in layers that can slide over each other. A hypothesis can be modified as new data is collected, and may even be disproved.

Data

Data can be displayed in **tables**, **pie charts** or **line graphs.** In your exam you may be asked to:

- choose the most appropriate method for displaying data
- identify trends
- use data mathematically – including using statistical methods, calculating the mean and calculating gradients of graphs.

A Table

% Yield	Temperature			
Pressure	250°C	350°C	450°C	550°C
200 atm	73%	50%	28%	13%
400 atm	77%	65%	45%	26%

A Pie Chart

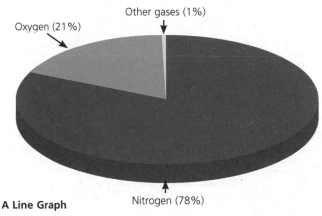

Oxygen (21%) Other gases (1%) Nitrogen (78%)

A Line Graph

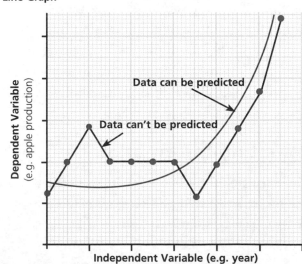

Data can be predicted

Data can't be predicted

Dependent Variable (e.g. apple production)

Independent Variable (e.g. year)

Fundamental Scientific Processes

Data (cont)

Sometimes the same data can lead to different conclusions. For example, data shows that the world's average temperatures have been rising significantly over the last 200 years. Some scientists think this is due to increased combustion of fossil fuels, while other scientists think it's a natural change that has happened before during Earth's history.

Scientific and Technological Development

Every scientific or technological development can have effects that we do not know about. This can give rise to **issues**. An issue is an important question that is in dispute and needs to be settled. Issues can be:

- **social** – they impact on the human population of a community, city, country or the world
- **environmental** – they impact on the planet, its natural ecosystems and resources
- **economic** – money and related factors such as employment and the distribution of resources
- **ethical** – what is right and wrong morally; a value judgement must be made
- **cultural** – giving an insight into differences between people on local and global scales.

Peer review is a process of self-regulation involving experts in a particular field who **critically examine** the work undertaken. Peer review methods are designed to maintain standards and provide **credibility** for the work that has been carried out. The methods used vary depending on the nature of the work and also on the overall purpose behind the review process.

Evaluating Information

Conclusions can then be made based on the scientific evidence that has been collected – they should try to explain the results and observations.

Evaluations look at the whole investigation. It is important to be able to evaluate information relating to social–scientific issues. **When evaluating information:**

- make a list of **pluses** (pros)
- make a list of **minuses** (cons)
- consider how each point might **impact on society**.

You also need to consider if the source of information is reliable and credible and to consider opinions, bias and weight of evidence.

Opinions are personal viewpoints – those backed up by valid and reliable evidence carry far more weight than those based on non-scientific ideas. Opinions of experts can also carry more weight than those of non-experts. Information is **biased** if it favours one particular viewpoint without providing a balanced account. Biased information might include incomplete evidence or it might try to influence how you interpret the evidence.

Examples of these processes are included within the main content of the book. However, it is important to remember that fundamental scientific processes are relevant to all areas of science.

B1: Understanding Organisms

This module looks at:
- Health and fitness, and problems caused by lifestyle choices.
- Human health and diet.
- Diseases and their causes, prevention and cures.
- How the nervous system allows the body to respond to changes in the environment.
- The risks and effects of drugs, alcohol and smoking.
- How processes take place in cells and organs to achieve a constant internal environment.
- Growth and development in plants, and how human intervention can affect this.
- People's characteristics and the causes of variation.

Health and Fitness

Health and fitness refer to a person's physical wellbeing. Being healthy primarily means being free from infectious disease, together with having total physical and mental wellbeing. Being fit relates to how much physical activity you are capable of doing and how quickly your body recovers afterwards.

Different types of exercise develop different aspects of fitness, which are all measurable – for example, strength, stamina, flexibility, agility and speed.

In your exam you might be asked to analyse these ways of measuring fitness.

> **HT** You may also be asked to evaluate these methods.

Cardiovascular efficiency – how well the heart copes with aerobic exercise and how quickly it recovers afterwards – is often used as a measure of general fitness.

Pulse

Each time the heart beats, the arteries **pulse** (throb). This can be felt at certain points on the body where the arteries are close to the surface, e.g. the wrist or neck.

Pulse rate is measured in beats per minute (**bpm**). A normal resting pulse rate for an adult is between 60bpm and 100bpm. Fitness is normally assessed by measuring how quickly an individual's pulse rate **recovers** after exercise, i.e. by timing how long it takes for the pulse rate to return to the normal resting pulse rate. The fitter a person is, the faster the recovery time.

The Circulatory System

The **heart** pumps blood around the body in **blood vessels** called **arteries**, **capillaries** and **veins**. It beats automatically, but the rate varies depending on the body's level of stress and exertion.

The heart is a **muscular** pump. It alternately relaxes to fill with blood and contracts to squeeze the blood out into the arteries, so the blood is always under **pressure**. This ensures that it reaches all the cells to supply them with oxygen and glucose for respiration. It also enables carbon dioxide to be removed by the lungs as a waste product.

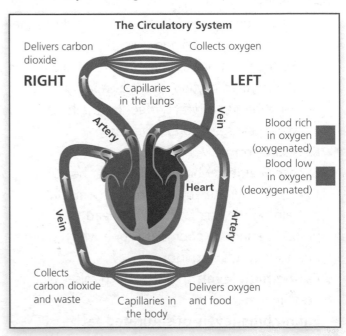

The Circulatory System

Delivers carbon dioxide

Collects oxygen

RIGHT   **LEFT**

Capillaries in the lungs

Artery   Vein

Blood rich in oxygen (oxygenated)

Blood low in oxygen (deoxygenated)

Heart

Vein   Artery

Collects carbon dioxide and waste

Capillaries in the body

Delivers oxygen and food

Blood Pressure

Blood pressure is a measure of **the force of blood per unit area** as it flows through the arteries. It is measured in mm Hg (millimetres of mercury).

Blood pressure is at its highest when the heart muscle contracts, forcing blood into the arteries to all parts of the body. This is called the **systolic blood pressure**. When the heart relaxes, the pressure in the arteries drops. This is called the **diastolic blood pressure**.

Healthy Blood Pressure

Normal blood pressure is about **120/80mmHg** (120mmHg is the systolic pressure; 80mmHg is the diastolic pressure). However, it can be affected by age and lifestyle.

Regular **aerobic exercise** strengthens the heart and helps maintain a normal blood pressure. A **healthy balanced diet** can also help.

An abnormally high blood pressure can increase the risk of heart disease. Obviously, if the factors above are controlled then they will cause a decrease in blood pressure.

Factors that increase blood pressure include:
- **High stress levels**.
- **Smoking**, which increases blood pressure in two ways:
  - The carbon monoxide it produces reduces the ability of the red blood cells to carry oxygen. So the heart has to beat faster to make up for the lowered oxygen-carrying capacity.
  - Nicotine has a direct effect on the heart, which causes it to beat faster.
- **Excess alcohol**.
- **Excess weight** (which can lead to obesity) due to lack of exercise and a poor diet. This puts a strain on the heart, which can lead to high blood pressure. A poor diet is one which is high in saturated fat, sugar or salt.
- **Eating high levels of salt**, which raises blood pressure.
- **Eating high levels of saturated fat**.

Saturated fats in the diet lead to a build-up of **cholesterol** (plaques) in arteries and restricts blood flow. The amount of cholesterol is linked to the amount of saturated fat eaten.

> **HT** Long-term high blood pressure is dangerous because the blood vessels can weaken and eventually burst. Burst blood vessels in the brain (called an aneurism) or in the kidneys can cause permanent damage.
>
> Narrowed coronary arteries, together with a **thrombosis**, increase the risk of heart attack or stroke.
>
> In carbon monoxide poisoning the gas combines with haemoglobin in red blood cells, preventing them from carrying as much oxygen.
>
> Low blood pressure, usually caused by weak pumping of the heart, can also be a serious problem. The blood does not circulate efficiently, so some parts of the body are deprived of glucose and oxygen. This can lead to dizziness and fainting, together with poor circulation which causes cold hands and feet.

You may be required to analyse data about the changing incidence of heart disease in a population.

You may be asked to interpret data showing possible links between the amount of saturated fat eaten, the build up of cholesterol plaques and the incidence of heart disease.

A Balanced Diet

Food is essential for all living organisms because it supplies them with the **energy** and nutrients they need in order to grow and function.

Recommended Servings per Day

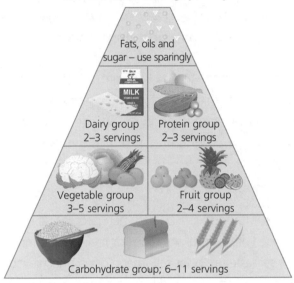

Fats, oils and sugar – use sparingly

Dairy group 2–3 servings | Protein group 2–3 servings

Vegetable group 3–5 servings | Fruit group 2–4 servings

Carbohydrate group; 6–11 servings

A **balanced diet** must contain:
- **carbohydrates** and **fats** to provide **energy**
- **protein** for **growth and repair of tissues** (and energy if fats and carbohydrates are unavailable).

Other substances are also needed in a balanced diet to keep us healthy, although they do not provide energy. They include:
- **minerals**, such as **iron**, which is needed to make haemoglobin in red blood cells
- **vitamins**, such as **vitamin C** which is needed to prevent scurvy
- **fibre**, which prevents **constipation**
- **water**, which prevents **dehydration**.

Everyone needs a balanced diet but people's diets vary greatly. This could be as a result of:
- beliefs about animal welfare, or their concerns about the effect of farming practices on food, e.g. vegetarians, vegans
- religious beliefs, e.g. the Muslim and Jewish faiths prohibit the eating of pig meat
- food allergies, e.g. people who are allergic to peanuts may suffer anaphylactic shock if they eat them.

How Much Energy is Needed?

The amount of energy needed by an individual depends on their age, sex and activity levels.

To maintain a healthy body mass, it is important to **balance** the amount of **energy consumed** in food with the amount of **energy expended** (used up) through daily activity.

Many people in the developed world do not get this balance right and become very overweight or **obese**. Obesity is a major health problem. It can lead to arthritis (swollen and painful joints), heart disease, diabetes and breast cancer.

Body Mass Index

One way to show whether someone is underweight or overweight for their height is to calculate their **body mass index** (**BMI**), using the following formula:

$$BMI = \frac{Mass \text{ (kg)}}{(Height \text{ in m})^2}$$

Recommended BMI Chart

BMI	What it Means
<18.5	Underweight – too light for your height
18.5–25	Ideal – correct weight range for your height
25–30	Overweight – too heavy for your height
30–40	Obese – much too heavy. Health risks!

Example
Calculate a man's BMI if he is 1.65m tall and weighs 68kg.

$$BMI = \frac{Mass \text{ (kg)}}{(Height \text{ in m})^2}$$
$$= \frac{68}{1.65^2} = \frac{68}{2.7} = 25$$

The recommended BMI for his height (1.65m) is 18.5–25, so he is a healthy weight.

Molecules in Food

Carbohydrates are made up of simple sugars such as glucose. **Fats** are made up of fatty acids and glycerol.

> **HT** Food molecules that are not used immediately are stored as different molecules. Carbohydrates are stored in the liver as glycogen or are converted to fat. Fats are stored under the skin and around organs as **adipose tissue**. Proteins are not stored.

Protein

Protein molecules are long chains of amino acids.

> **HT** There are different types of amino acids:
> - **essential amino acids** must be taken in by eating food (your body cannot make them).
> - **non-essential amino acids** can be made in the body.
>
> Your diet should include the complete range of amino acids so that your body can make all the necessary proteins. Meat and fish are **first class proteins** because they contain all the different types of amino acids. Plant proteins are called **second class proteins**.
>
> Very few vegetables contain all the necessary amino acids, so a vegetarian or vegan diet must include a variety of plant protein (especially beans and pulses) so that all the amino acids are included.

A diet that does not contain enough protein will not allow normal growth. This is why it is important for teenagers to have a high-protein diet.

In some parts of the world, food is in very short supply. This may be due to overpopulation and limited investment in agricultural techniques. Often the only source of food is a cereal like rice. Rice is not a good source of protein. In children, protein deficiency results in a disease called **kwashiorkor**. The muscles waste because the proteins in them are used for energy, and the stomach swells due to too much fluid.

How Much Protein?

The **estimated average requirement (EAR)** of protein intake is calculated using the formula:

$$\text{EAR protein (g)} = 0.6 \times \text{Body mass (kg)}$$

How much protein would a woman weighing 60kg need?
EAR $= 0.6 \times 60 =$ **36g protein per day**

> **HT** EAR is an estimated daily figure for an average person of a certain body mass. The EAR for protein may vary depending on age, or whether a woman is pregnant or producing milk.

Eating Disorders

Problems with food can begin when people use it to help cope with painful situations or feelings. Feelings of low self-esteem, poor self-image or an unachievable desire for perfection can all lead to a poor diet. People might start using food as a source of comfort, or restricting what they eat to lose weight or because it makes them feel more in control of their lives. When food is used in this way, it is called an **eating disorder**. Eating disorders, and even some diets, are damaging because the body is no longer getting the right balance of energy and nutrients it needs to function properly.

People who have **anorexia nervosa** restrict what they eat, and sometimes starve themselves. This can lead to extreme weight loss and poor growth, constipation and abdominal pains, dizzy spells and feeling faint, a bloated stomach, poor circulation, discoloured skin, irregular or no periods (in girls) and loss of bone mass which can eventually develop into osteoporosis (brittle bones).

People who suffer from **bulimia nervosa** make themselves vomit, or take laxatives after eating, to get the food out of their system before it can be digested. This can lead to large weight fluctuations, a sore throat, tooth decay and bad breath, swollen salivary glands, poor skin condition and hair loss, irregular periods (in girls), tiredness and an increased risk of problems with the heart and other internal organs.

Disease

Infectious diseases are caused by **microorganisms** which attack and invade the body. They are spread from one person to another through unhygienic conditions or contact with an infected person.

Non-infectious diseases cannot be caught from another person. They can be caused by:

- **poor diet**, e.g. a lack of vitamin C causes scurvy, and a lack of iron causes anaemia
- **organ malfunction**, e.g. the pancreas can stop producing insulin (which causes diabetes), or cells can mutate and become cancerous
- **genetic inheritance**, e.g. people can inherit the genes for a particular condition or disorder from their parent, such as red–green colour blindness or cystic fibrosis.

You may be required to look for connections between the incidence of disease and things such as climate factors and socio-economic factors.

Cancer

Cancer is a non-infectious disease caused by **mutations** in living cells. Making healthy lifestyle choices is one way to reduce the likelihood of cancer, for example:

- do not smoke – smoking causes lung cancer
- do not drink excess alcohol – alcohol consumption is linked to cancer of the liver, gut and mouth
- avoid getting sunburnt – skin cells damaged by the Sun can become cancerous
- eat a healthy diet – a high-fibre diet can reduce the risk of bowel cancer.

HT Cancerous cells divide in an abnormal and uncontrolled way, forming lumps of cells called **tumours**. If a tumour grows in one place it is described as **benign**. However, if cells break off and secondary tumours start to grow in other parts of the body, they are described as **malignant**.

A person's **chance of survival** depends on the type of cancer they have. People with **breast** or **prostate** cancer have high survival rates, whereas people with **lung** or **stomach** cancer have very low survival rates. The chance of survival is greater if the cancer is diagnosed early and the patient is young.

Pathogens

Pathogens are disease-causing microorganisms. The different types are listed below:

- **fungi**, e.g. athlete's foot
- **viruses**, e.g. flu
- **bacteria**, e.g. cholera
- **protozoa**, e.g. malaria, amoebic dysentery.

Malaria

Malaria is a disease caused by a **protozoan**, which is a **parasite**. Parasites live off other organisms, called **hosts**. In the case of malaria, humans are the hosts.

Malaria parasites can be sucked up from a human's bloodstream by mosquitoes (a **vector**). Once inside the mosquito, they mate and move from the gut to the salivary glands. When the mosquito bites another person, the malaria parasites are passed on into their bloodstream. They then head straight for the liver, where they mature and reproduce. The malaria parasites then migrate to the blood and replicate in red blood cells, bursting them open. This damage leads to characteristic malaria fever and can sometimes result in death.

The parasite which causes malaria reproduces and matures much more quickly in warmer climates, and the mosquito vectors can only breed in stagnant water. This is why the pattern of malarial diseases is linked to areas with warm, wet conditions.

HT The best way to prevent a disease like malaria spreading is to control the vectors. In countries where malaria is common, people sleep under mosquito nets and use insect repellents to avoid getting bitten by mosquitoes. Mosquitoes can be killed by spraying them with insecticide.

The most effective ways of controlling malarial mosquito numbers is to target specific points in their life cycle. For example, the larval stage can be targeted by draining stagnant water or spraying these areas with oil or pesticide.

B1 | Staying Healthy

Defences Against Pathogens

The body has a number of defences to stop pathogens getting in:

- The **skin** acts as a barrier against microorganisms.
- The **respiratory system** is lined with specialised cells that produce a **sticky**, **liquid mucus** that forms a **mucus membrane** which traps microorganisms. Tiny hairs called **cilia** move the mucus up to the mouth where it is swallowed.
- The **stomach** produces hydrochloric acid which kills microorganisms on the food we eat.
- The **blood** clots in wounds to prevent microorganisms from entering the bloodstream.

Dealing with Pathogens Inside the Body

If pathogens manage to enter the body, the white blood cells start reacting to the invasion. (The symptoms of a disease are caused by pathogens **damaging cells** and **producing toxins (poisons)** before the white blood cells can destroy them.)

Phagocytes are a type of white blood cell that move around in the bloodstream searching for pathogens. When they find some, they **engulf** and **digest** them. When we get an infected cut and pus develops, the yellow liquid is mainly white blood cells full of digested microorganisms.

Lymphocytes are another type of white blood cell. They recognise molecular markers called **antigens** on the surface of the pathogen and produce **antibodies** that lock onto the antigens and kill the pathogens:

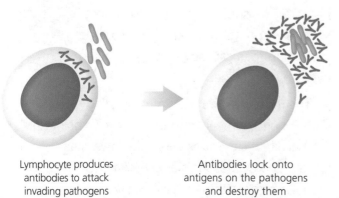

Lymphocyte produces antibodies to attack invading pathogens

Antibodies lock onto antigens on the pathogens and destroy them

HT Every pathogen has its own unique antigens. White blood cells make antibodies specifically for a particular antigen. So, for example, antibodies made to react to tetanus have no effect on whooping cough or cholera.

Active and Passive Immunity

Once white blood cells are sensitised to a particular pathogen, they can produce the necessary antibodies much quicker if the same pathogen is detected again. This provides future protection against the disease and is called **active immunity**. Active immunity can also be achieved through **vaccination** (**immunisation**).

Passive immunity occurs when antibodies are put into an individual's body, rather than the body producing them itself. For example, some pathogens or toxins (e.g. snake venom) act very quickly, and a person's immune system cannot produce antibodies to destroy the pathogen quickly enough. Therefore, the person must be injected with the antibodies. However, they will not have long-term protection against the pathogen because their white blood cells did not produce the antibody themselves.

HT Immunisation

Immunisation provides immunity to a disease, without the person being infected by it, as follows:

1. A weakened or dead strain of the pathogen is injected. The pathogen is heat-treated so it cannot multiply inside the person. But the antigen molecules remain intact.
2. Even though they are harmless, the antigens on the pathogen trigger the white blood cells to produce specific antibodies.
3. The white blood cells, called **memory cells**, remain 'sensitised', which means they can produce more antibodies very quickly if the same pathogen is detected again.

HT

Benefits
• It protects against diseases that could kill or cause disability, e.g. polio, measles. • If everybody is vaccinated, the disease cannot spread and eventually dies out. (This is what happened to smallpox.)
Risk
• An individual could have an allergic reaction to the vaccine.

Some research has linked the **MMR** (measles, mumps and rubella) vaccine to children developing bowel problems and autism. As a result, lots of parents decided not to vaccinate their children and the number of cases of measles then immediately began to rise. However, much of this research has since been discredited.

Treating Diseases with Drugs

Diseases caused by bacteria or fungi (not viruses) can be treated using **antibiotics**. These are drugs that kill the pathogen. Viral diseases can be treated with antiviral drugs, e.g. flu can be alleviated by taking 'Tamiflu' tablets.

Antibiotics are very effective at killing bacteria. However, there are some bacteria that are **naturally resistant** to particular antibiotics.

It is important for patients to follow instructions carefully and take the full course of antibiotics to give them the best possible chance of killing all the bacteria. If doctors over-prescribe antibiotics, all the bacteria in a population are killed off except the resistant ones, which will then spread. The antibiotic then becomes useless. **MRSA** is a bacterium which has become resistant to most antibiotics, making it a dangerous microorganism that the media has dubbed a 'superbug'.

Drug Testing

New drugs have to be developed all the time to combat different diseases. The drugs must be tested to make sure that they are **effective** and **safe**. A drug can be tested using:

- **computer models** to predict how it will affect cells, based on knowledge about how the body works and the effects of similar drugs
- **animals** to see how it affects living organisms (some people object to this on the grounds of animal cruelty)
- **human tissue** (grown in a laboratory) to see how it affects human cells (some people object to human tissue being grown in this way because they believe it is unnatural and wrong).

Finally, the drug must be tested on healthy volunteers, and on volunteers who have the relevant disease. Some of them are given the new drug and some are given a **placebo** (an inactive pill). The effects of the drug can then be compared to the effects of taking the placebo.

In a **blind trial** the volunteers do not know whether they have been given the new drug or the placebo. This eliminates any psychological factors and helps to provide a **fair comparison**.

In a **double blind trial** neither the volunteers nor the doctors know which pill has been given. This eliminates **all** bias from the test, because the doctors cannot influence the volunteers' response in any way.

New drugs must also be tested against the best existing treatments.

Although scientists conduct lots of tests beforehand to determine how the drugs will affect humans, drug trials like these can never be completely safe.

Nerve Cells (Neurones)

Neurones are **specially adapted cells** that can carry a **nerve impulse**, which is electrical in nature. This is carried in the **axon** (the long, thin part). There are three types of neurone:

1. A **sensory neurone** carries nerve impulses from the receptors to the central nervous system.
2. A **relay neurone** makes connections between neurones inside the brain and spinal cord.
3. A **motor neurone** carries nerve impulses from the central nervous system to the muscles and glands.

A Motor Neurone

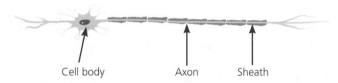

Cell body   Axon   Sheath

HT Motor neurones are specially adapted to carry out their function. They have:

- an **elongated** (long) shape to make connections from one part of the body to another
- an insulating **sheath**, which speeds up the nerve impulse
- **branched endings** (dendrites), which allow a single neurone to connect to many other neurones.

There is a small gap between neurones called a **synapse**. An electrical impulse travels down a neurone until it reaches a synapse. A transmitter substance is then released across the synapse and it travels by diffusion. The transmitter binds with **receptor molecules** on the next neurone, causing an electrical impulse to be released in that neurone.

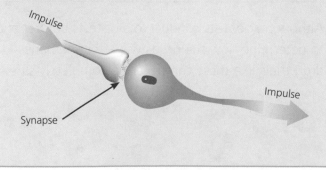

Impulse

Impulse

Synapse

The Nervous System

The nervous system allows organisms to **react** to their surroundings and **coordinate** their behaviour. The nervous system can be divided into the **central nervous system** and the **peripheral nervous system**.

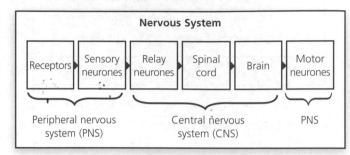

Nervous System

Receptors → Sensory neurones → Relay neurones → Spinal cord → Brain → Motor neurones

Peripheral nervous system (PNS)   Central nervous system (CNS)   PNS

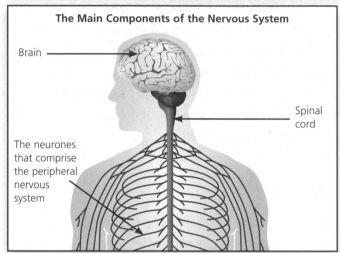

The Main Components of the Nervous System

Brain

Spinal cord

The neurones that comprise the peripheral nervous system

Receptors

Animals detect changes in their environment, called **stimuli**, using **receptors**, which generate nerve impulses. Receptors are specialised nerve endings. The different types are:

- light receptors in the retina of the eye – detect light
- sound and change of position receptors in the ears (for balance)
- taste receptors on the tongue – detect chemicals in food
- smell receptors in the nose – detect chemicals in the air
- touch, pressure, pain and temperature receptors in the skin.

Effectors are the **muscles** or **glands** that make the change in response to the signal from the receptor.

Voluntary Action

Voluntary responses are under the **conscious control** of the **brain**, i.e. the person **decides** how to react to a stimulus. The pathway for processing the information and acting on it is:

stimulus ➔ receptor ➔ sensory neurone ➔ coordinator ➔ motor neurone ➔ effector ➔ response

Example

1. **Pressure receptors** in the skin of your thigh detect an insect crawling on you.
2. The receptors cause an **impulse** to travel along a **sensory neurone**.
3. The impulse is received in the **spinal cord**. Here, another sensory neurone carries the impulse to the **brain** (coordinator).
4. The brain thinks about the impulse and decides to flick the insect away with the left hand.
5. An impulse is sent down a **motor neurone** in the spinal cord.
6. This causes an impulse to be sent out of the spinal cord via another motor neurone.
7. The impulse is received by a muscle (an **effector**) in the left hand, causing the hand to move and flick away the insect (a **response**).

Reflex Action

Reflex actions (involuntary responses) give **fast**, **automatic responses** to a stimulus, helping to protect the body from harm. The pathway for receiving and acting on information is called the **reflex arc**:

stimulus ➔ receptor ➔ sensory neurone ➔ relay neurone ➔ motor neurone ➔ effector ➔ response

Some useful reflex actions include:
- eye pupil reflex automatically controls the amount of light that enters the eye (prevents damage to retina)
- knee-jerk reaction
- withdrawing your hand from a hot plate to prevent you from getting burned.

Example

1. A **receptor** is stimulated by the pin (stimulus).
2. This causes impulses to pass along a **sensory neurone** into the **spinal cord** (the coordinator in reflex actions).
3. The sensory neurone synapses with a **relay neurone**, **bypassing the brain**.
4. The relay neurone synapses with a **motor neurone**, sending impulses down it.
5. The impulses are received by the muscles (**effectors**), causing them to contract and remove the hand in **response** to the pain.

Example of Reflex Action (Reflex Arc)

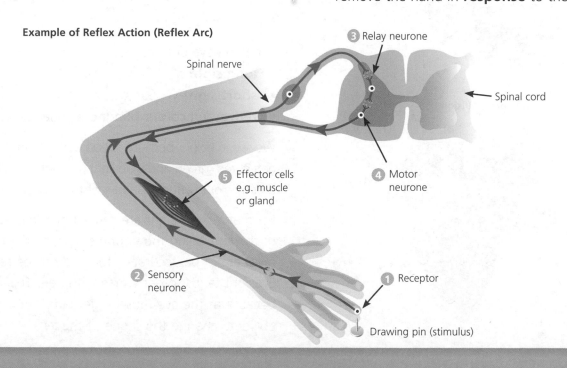

- Spinal nerve
- 3 Relay neurone
- Spinal cord
- 5 Effector cells e.g. muscle or gland
- 4 Motor neurone
- 2 Sensory neurone
- 1 Receptor
- Drawing pin (stimulus)

Vision

Humans and some animals have eyes positioned close together on the front of their head. This is called **binocular vision**, and is usually found in predators. Each eye has a narrow or limited field of view but, where the fields of view overlap, the brain interprets the information and creates a 3-D image. This means that the animal can judge distance and speed quite accurately.

Some animals have eyes set on each side of their head. This is called **monocular vision**, and is usually found in prey. Each eye has a wide field of view, so as well as being able to see to each side, the animal can also see behind and in front. However, there is very little overlap in the fields of view so it is difficult for the animal to judge distance or speed.

Binocular vision helps to judge distances by comparing the images from each eye. The more similar the images, the further away the object.

Binocular Vision

Monocular Vision

The Eye

The eye focuses light onto the retina. The **cornea** and **lens refract** rays of light, so they converge (come together) at a single (focal) point on the **retina**. This stimulates the light-sensitive receptor cells in the retina and causes nerve impulses to pass along sensory neurones to the brain. The brain then interprets the impulses, a process sometimes known as perception.

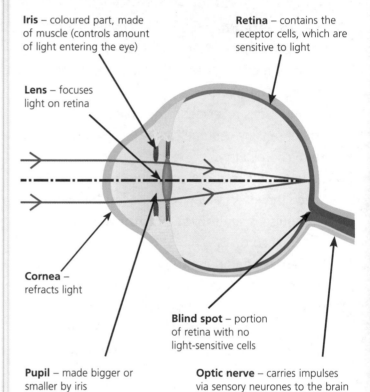

Iris – coloured part, made of muscle (controls amount of light entering the eye)

Retina – contains the receptor cells, which are sensitive to light

Lens – focuses light on retina

Cornea – refracts light

Blind spot – portion of retina with no light-sensitive cells

Pupil – made bigger or smaller by iris

Optic nerve – carries impulses via sensory neurones to the brain

Eye Defects

Common eye defects are:
- **long sight**
- **short sight**
- **red–green colour blindness** (inherited condition).

We see in colour because specialised cells in the retina detect red, green and blue light. In people with red-green colour blindness, some of these cells are missing.

Long and short sight are caused by the eyeball or the lens being the wrong shape, so the light rays cannot be accurately focused on the retina. Eyesight tends to get worse with age. This is because as the eye muscles get older they lose the ability to change the shape of the lens.

More about the Eye

The **lens** is a clear, flexible bag of fluid surrounded by circular ciliary muscles that change the shape of the lens (accommodation). **Suspensory ligaments** attach the lens to the ciliary muscles.

Light rays reflected by a **distant** object are almost **parallel** when they reach the eye, so:
- the ciliary muscles **relax**
- the suspensory ligaments become **taut**
- the lens becomes **long and thin**
- light is only **refracted a little** to focus on the retina.

Light rays reflected by a **near** object **diverge** (are reflected out in all directions), so:
- the ciliary muscles **contract**
- the suspensory ligaments go **slack**
- the lens becomes **short and fat**
- light is **refracted** to focus on the retina.

Section view

Focus on a distant object   Focus on a near object

Front view

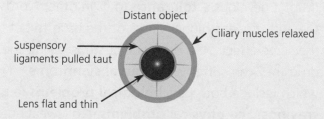

Distant object

Suspensory ligaments pulled taut

Ciliary muscles relaxed

Lens flat and thin

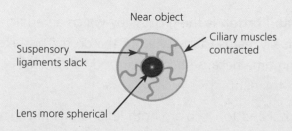

Near object

Suspensory ligaments slack

Ciliary muscles contracted

Lens more spherical

Long Sight

Long sight is caused by an eyeball that is too short, or by a lens that stays long and thin (i.e. does not change shape properly). It can be corrected by a **convex** lens which **converges** the light rays from close objects so that they focus on the retina.

Long sightedness is when you can see distant objects clearly but not close objects.

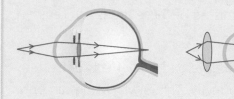

Short Sight

Short sight is caused by an eyeball that is too long, or weak suspensory ligaments which cannot pull the lens into a thin, flat shape. It can be corrected by a **concave** lens which **diverges** the light rays so that they focus on the retina.

Short sightedness is when you can see near objects clearly but not distant objects.

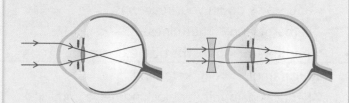

Corrective lenses can be worn either as contact lenses or glasses. Some people opt for laser surgery to change the shape of the cornea and lens.

Drugs and You

Drugs

Drugs are chemicals that affect the way the body works. They can affect the mind or the body (or both) and are used for **medicine** and **pleasure** (recreational drugs). Some drugs used for pleasure are legal, e.g. tobacco and alcohol, whereas others are illegal, e.g. ecstasy and cocaine.

Illegal drugs are **harmful**. They can have very bad side effects and can lead to **addiction** and even **death**.

Some drugs, e.g. Valium, can only be prescribed by a doctor because misuse of these drugs can cause harm.

Even medicines can have bad side effects or be deadly if used incorrectly. This is why some can only be obtained on prescription. The effects of some types of drugs are briefly described below:

- **Stimulants**, e.g. caffeine, nicotine, ecstasy:
  - Increased brain activity helps to combat depression.
  - Increased alertness and perception.
  - Caffeine and nicotine can be addictive.
- **Depressants or sedatives**, e.g. alcohol, solvents, tranquillisers (like temazepam):
  - Decreased brain activity makes you feel tired and slows down your reactions.
  - Lethargy and forgetfulness.
- **Painkillers or anaesthetics**, e.g. aspirin, paracetamol, heroin, ketamine:
  - Reduces pain felt (by blocking nerve impulses).
  - Can be very addictive, especially heroin.
- **Performance-enhancing drugs**, e.g. anabolic steroids:
  - Increased muscle development (sometimes abused by sports people).
- **Hallucinogens**, e.g. cannabis, LSD:
  - Distorted perceptions, sensations and emotions.

Drug Classification

In the UK, illegal drugs are **classified** into three main categories under the **Misuse of Drugs Act**:

- **Class A** drugs, e.g. heroin, cocaine, ecstasy, and LSD are deemed the most dangerous and carry heavy prison sentences and fines for possession.
- **Class B** drugs, e.g. amphetamines, speed, cannabis and barbiturates.
- **Class C** drugs, e.g. tranquilisers and anabolic steroids, are less dangerous and have lower penalties.

Stimulants and Depressants

Stimulants increase the amount of **transmitter substance** released at synapses in the nervous system. This increases the level of nervous activity taking place in the brain, giving rise to feelings of energy, alertness and euphoria.

Depressants bind with receptor molecules in the membrane of the next neurone, blocking the transmission of the impulses. This reduces brain activity, slowing down reactions and making a person feel dopey, subdued or drowsy.

Addiction and Rehabilitation

An **addiction** is a **psychological need** for something, which means you always want more. As an addict's body becomes more used to the drug, it develops a **tolerance** to it. In other words, the addict needs higher doses of the drug to get the same effects. If a drug addict stops taking the drug they can suffer **withdrawal symptoms**, which include both **psychological problems** (e.g. **cravings**) and **physical problems** (e.g. sweating, shaking, nausea).

Rehabilitation is the process by which an addict gradually learns to live without the drug. This takes a long time because both their body and mind have to adapt.

Alcohol and tobacco are both legal drugs but they can still have serious effects on health.

Alcohol

Alcohol contains **ethanol**, which is a **depressant** and causes **slow reactions**. The **short-term effects** on the brain and nervous system can lead to lack of balance and muscle control, blurred vision, slurred speech, drowsiness, poor judgement and vasodilation (the widening of blood vessels in the skin, increasing blood flow and heat loss). **Excess** alcohol can lead to **unconsciousness** and **coma** or **death**. The **long-term effects** of alcohol can be liver damage or **brain damage** (due to dehydration).

> **HT** Cirrhosis (liver damage due to the liver having to work hard to remove the toxic alcohol from the body) occurs because toxic products build up in the liver when enzymes break down alcohol.

One **unit** of alcohol is **10cm³** of pure alcohol. The maximum advised weekly intake for men is 21 units and for women it is 14 units.

- Pint of strong lager – 3 units.
- Pint of ordinary strength lager, bitter or cider, or a large glass (175cm³) of wine – 2 units.
- Pub measure of spirits – 1 unit.

The **legal limit** for driving after drinking is 80 milligrams of **alcohol** in 100 millilitres of blood (80mg/100ml). This cannot easily be converted into units because the effects of alcohol depend on many factors including age, sex, height and how much food has been eaten. This legal limit has been set because alcohol slows reaction times and increases the chance of accidents. It is important to remember that even one alcoholic drink affects your ability to react, so it is better not to drink anything at all before driving. Similarly, pilots of aircraft have legal limits on their blood alcohol levels (20mg/100ml).

Tobacco

Tobacco smoke contains tar, carbon monoxide, particulates and **nicotine**, which is very **addictive**. Some of these chemicals damage the **cilia** (ciliated epithelial cells) that line the airways (trachea, bronchi, bronchioles), so they cannot remove the mucus, tar and dirt from the lungs. This leads to a **smoker's cough**. Excess coughing can damage the alveoli and cause **emphysema**.

Tar contains chemicals that are irritants and carcinogens (cancer-causing chemicals). Particulates from the cigarette accumulate in living tissue and this can cause mouth, throat, lung and oesophagus **cancer**. Other chemicals in cigarette smoke cause **high blood pressure** and increase the chance of **blocked blood vessels**, which cause **strokes** and **heart disease**. Particulates in cigarette smoke can accumulate in lung tissue. **Bronchitis** is common in smokers because the mucus in their lungs leaves them susceptible to infection, and they are often breathless because their cells are not getting enough oxygen – the haemoglobin in their red blood cells pick up **carbon monoxide** instead of **oxygen**.

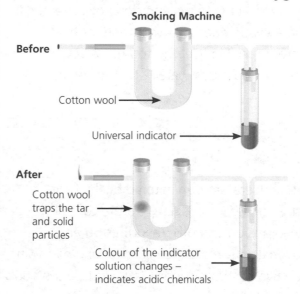

Smoking Machine

Before

Cotton wool

Universal indicator

After

Cotton wool traps the tar and solid particles

Colour of the indicator solution changes – indicates acidic chemicals

Carbon monoxide and nicotine are gases that cannot be detected in a simple smoking machine like this one, but are analysed by more sophisticated detectors. Carbon monoxide encourages fat deposits in arteries, which leads to **arteriosclerosis**. The arteries around the heart become blocked, which can lead to heart disease and, ultimately, a heart attack.

> **HT** You may be asked in your exam to evaluate data on the effects of smoking, for example rates of cancer, heart disease or emphysema or birth weights of babies born to mothers who smoke (they often have lower **birth weights**).

Homeostasis

The body has automatic control systems to maintain a constant internal environment (**homeostasis**) to ensure that cells can function efficiently. It balances inputs and outputs and removes waste products to ensure that the optimum levels of temperature, pH, water, oxygen and carbon dioxide are maintained. Controlling these factors is essential to life.

Since **enzymes** work best at **37°C** (the core temperature in humans), it is essential that the body remains very close to this temperature. Heat produced through respiration is used to maintain the body temperature.

If body temperature becomes too high, the blood flows closer to the skin so heat can be transferred to the environment. This is also done by increased sweating – evaporation of sweat requires heat energy from the skin. Getting too hot is very dangerous. If too much water is lost through sweating, the body becomes **dehydrated**. This can lead to **heat stroke** and even death.

If the body temperature falls too low, the blood flow near the skin is reduced, sweating is reduced and muscles start making tiny contractions, commonly known as shivers. These contractions need energy from respiration and heat is released as a by-product. Getting too cold can also be fatal. **Hypothermia** is when the body temperature drops too far below 37°C. This causes unconsciousness and sometimes death. If you start to feel cold, you should put on some more clothing and do some exercise.

Body temperature readings can be taken from the mouth, ear, skin surface, finger or rectum. Although an anal temperature reading is the most accurate, it is normally only used in hospitals. Digital recording probes and thermal imaging are also used in hospitals. At home, heat-sensitive strips that are placed on the forehead are an alternative to thermometers.

Vasodilation and Vasoconstriction

Blood temperature is monitored by the brain, which switches various temperature control mechanisms on and off. **Vasodilation** and **vasoconstriction** are the widening and narrowing (respectively) of the blood vessels close to the skin's surface to increase or reduce heat loss.

Negative Feedback

Negative feedback occurs frequently in homeostasis. It involves the automatic reversal of a change in condition. For example, if the temperature falls too low, the brain switches on mechanisms to increase it. If the temperature then becomes too high, the brain switches on mechanisms to lower it. Just like a central heating thermostat! The brain exerts control through nervous and hormonal systems.

Core body temperature too high	Thermoregulatory Centre	Core body temperature too low

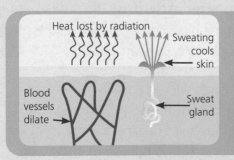

Vasodilation

Heat lost by radiation
Sweating cools skin
Blood vessels dilate
Sweat gland

Body temperature needs to decrease, so blood vessels in the skin dilate (become wider) causing greater heat loss, as more heat is lost from the surface of the skin by radiation.

Vasoconstriction

Sweating stops
Blood vessels constrict
Sweat gland

Body temperature needs to increase – blood vessels in skin constrict (become narrower) reducing heat loss, as less heat is lost from the surface of the skin by radiation.

Controlling Blood Sugar

The **pancreas** is an endocrine organ located in the upper abdomen. It produces a **hormone** called **insulin**, which travels around the body in the blood.

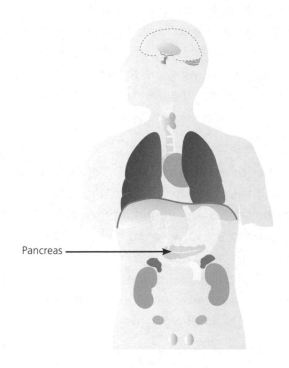

Pancreas

In general, hormones act more slowly than nervous responses because it takes more time for these chemicals to be transported in the bloodstream.

Diabetes

Diabetes is caused by the **pancreas** not producing enough of the hormone **insulin**, which controls blood sugar levels. This can lead to blood sugar levels rising fatally **high** and resulting in a **coma**, so blood sugar has to be **controlled** by **injecting** insulin or controlling the amount of **sugar** in the diet.

Type I and Type II Diabetes

Type I diabetes is more likely to occur in the under 40s and is the most common type in childhood. It is caused by failure of the pancreas to produce insulin. This is often the result of an auto-immune response where cells in the pancreas are destroyed.

Type II diabetes accounts for 85% of all cases and is most common in the over 40s in the white population and in the over 25s in the black and south Asian population. The cause is under-production of insulin or the inability of the hormone to interact with body cells, often a result of fatty deposits in obese people.

Type I diabetes usually has to be controlled by insulin injections, whereas type II can often be controlled by diet.

HT Insulin helps to regulate blood sugar levels by converting excess **glucose** in the blood to **glycogen** in the liver.

Before injecting insulin, a person with Type I diabetes tests the amount of sugar in their blood. If they have had food containing a lot of sugar, then a bigger dose of insulin is required to reduce the blood sugar level. If they intend to exercise, then a smaller dose is required as they will use up a lot of sugar (for energy).

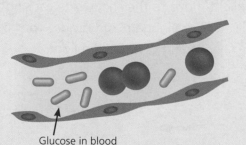

Glucose in blood

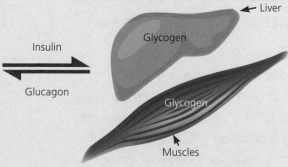

Insulin

Glucagon

Liver

Glycogen

Glycogen

Muscles

Plant Hormones

Plants, as well as animals, respond to changes in their environment. **Plant hormones (auxins)** are chemicals that control the **growth** of shoots and roots, **flowering** and the **ripening** of fruits. The different hormones move through the plant in solution and affect its growth by responding to **gravity (geotropism)** and **light (phototropism)**.

Shoots grow towards light (positive phototropism) and against gravity (negative geotropism). Growth towards the light can increase a plant's chance of survival because light is fundamental to photosynthesis. **Roots** grow away from light (negative phototropism) and in the direction of gravity (positive geotropism) for anchorage (support) and to absorb water and minerals.

An experiment can be carried out to show that shoots grow towards light. A hole is cut out of the side of two boxes to enable light to enter. Three cuttings of a plant are put in one of the boxes (see ❶). Three cuttings are put in the other box, but the tips of the shoots are covered with foil (see ❷). The normal shoot tips detect the light and grow towards it. However, the shoots covered in foil cannot detect the light so they grow straight up.

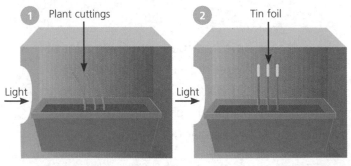

Commercial Uses of Hormones

Plant hormones can be used in agriculture:

- **Rooting powder** consists of a hormone that encourages the growth of roots in stem cuttings so lots of plants can be obtained from one plant.
- **Fruit-ripening hormone** causes fruit to ripen. Fruits are sometimes harvested while still under-ripe and sprayed with the hormone during transportation so that they are ripe when they reach the shops. Alternatively, they can be sprayed with a hormone that delays ripening.
- **Selective weedkillers** contain hormones that disrupt the growth patterns of their target plants without harming other plants.
- **Dormancy** can be controlled by hormones, to speed up or slow down plant growth and bud development, allowing the farmer to control dormancy and get the best price at market.

HT Gravity

Hormones called **auxins** travel in solution down towards the lower side of shoots and roots:

- In the shoots, the hormones increase growth in the lower region, which makes the shoot bend upwards.
- In the roots, the hormones slow down growth in the lower region, which makes the root bend downwards.

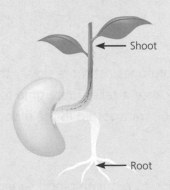

Light

Auxin is made in the shoot tip. Its distribution through the plant is determined by light. When light shines on a shoot, the hormones in direct sunlight move to the shady side causing the cells here to elongate and divide. This makes the shoot bend towards the light.

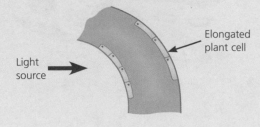

Genetic Information

The nuclei of living cells contain **chromosomes**, which are made up of a string of genes. A chromosome is a single piece of **coiled DNA** containing many genes. Different **genes** control the development of different characteristics by issuing instructions to cells.

Chromosomes are long lengths of **DNA** (deoxyribonucleic acid). The DNA molecule itself consists of **two strands** which are coiled to form a **double helix**.

The DNA molecules in a cell form a complete set of **instructions** on how the organism should be constructed and how its individual cells should work.

The instructions are in the form of a **chemical code** made up of four **bases**, which hold the two strands of the molecule together.

The order (or sequence) of the bases in a particular **region** of DNA (a gene) provides the **genetic code** that controls cell activity and the development of the related characteristic.

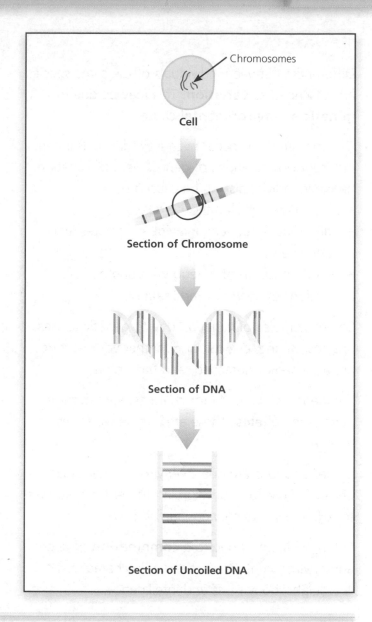

Chromosomes

Cell

Section of Chromosome

Section of DNA

Section of Uncoiled DNA

Reproduction

Most body cells have the **same number** of **chromosomes**, in matching **pairs** – humans have 23 pairs (46 chromosomes in total). Other species have different numbers of chromosomes, e.g. a chicken has 78 chromosomes.

However, the **gametes** (sex cells) contain individual chromosomes and therefore have exactly half the number of normal cells – 23 chromosomes in humans.

During gamete production, chromosome combinations in each cell are mixed up so that no two gametes are exactly the same. This is a source of **variation**. During **sexual reproduction**, when the male (sperm) and female (egg) gametes **fuse**, they produce a cell with the correct number of chromosomes. This process increases the amount of **variation** in a species because genetic information from two parents is mixed together to produce a unique new individual.

In **asexual reproduction** all the genes come from one parent. The offspring, which are genetically identical to the parent and to each other, are called **clones**.

Variation

Differences between individuals of the same species are described as **variation**. This may be due to **genetic** or **environmental** causes.

Genetic variations occur because individuals inherit different combinations of genes. Genetic variation between individuals can be caused by:

- mutations which alter the genes
- differences between individual gametes (eggs and sperm)
- random fusion of an egg with one out of millions of sperm at fertilisation.

Some examples of variation due to genetic causes are nose shape, eye colour, whether your earlobes are attached or detached, and hair colour.

The genes which code for characteristics occur in pairs called **alleles**. These are alternative forms of a gene.

Some variations are due to environmental causes, because individuals develop in different conditions – for example, spoken language and scars.

Often, variation is due to a **combination** of genetic and environmental causes – e.g. in characteristics like height, body mass and intelligence.

> **HT** Scientists are currently debating whether genetics or environment has the greatest influence in the development of characteristics like intelligence, health and sporting ability. It is unlikely that any characteristics are the sole result of one factor.

Inheritance of Sex

Sex (in mammals) is inherited via the **sex chromosomes**: one of the 23 pairs present in humans. The chromosomes are labelled **X** or **Y**. It is the presence of the Y chromosome that determines the sex of an individual: **XX = female; XY = male**.

HT All egg cells carry X chromosomes. Half the sperm carry X chromosomes and half carry Y chromosomes. The sex of an individual depends on whether the egg is fertilised by an X-carrying sperm or a Y-carrying sperm. If an X sperm fertilises the egg it will become a girl. If a Y sperm fertilises the egg it will become a boy. The chances of these events are equal, which results in approximately equal numbers of male and female offspring.

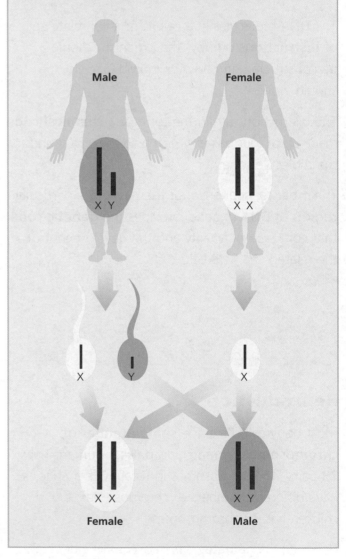

The Terminology of Inheritance

Genes can have different **alleles** (alternative versions). For example, the gene for eye colour has two alleles: brown and blue. Similarly, the gene for tongue rolling has two alleles: being able to roll your tongue and not being able to roll your tongue.

Alleles are described as being **dominant** or **recessive**:

- A **dominant** allele controls the development of a characteristic even if it is present on only one chromosome in a pair.
- A **recessive** allele controls the development of a characteristic only if a dominant allele is not present, i.e. if the recessive allele is present on both chromosomes in a pair.

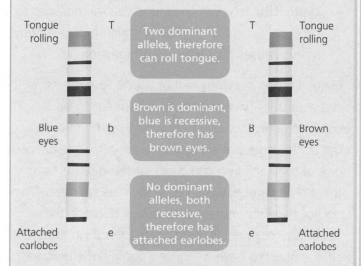

Tongue rolling — T — Two dominant alleles, therefore can roll tongue. — T — Tongue rolling

Blue eyes — b — Brown is dominant, blue is recessive, therefore has brown eyes. — B — Brown eyes

Attached earlobes — e — No dominant alleles, both recessive, therefore has attached earlobes. — e — Attached earlobes

N.B. A capital letter is used for a dominant allele, and a small letter is used for a recessive allele.

If **both chromosomes** in a pair contain the **same allele** of a gene, the individual is described as being **homozygous** for that gene or condition.

If the chromosomes in a pair contain **different alleles** of a gene, the individual is **heterozygous** for that gene or condition.

The combination of alleles for a particular characteristic is called the **genotype**. For example, the genotype for a homozygous dominant tongue-roller would be **TT**. The fact that this individual is able to roll their tongue is termed their **phenotype**. Other examples are **bb** (genotype), blue eyes (phenotype); **Ee** (genotype), and unattached ear lobes (phenotype).

When a characteristic is determined by just one pair of alleles, as with eye colour and tongue rolling, it is called **monohybrid inheritance**.

Genetic Diagrams

Genetic diagrams are used to show all the possible combinations of alleles and outcomes for a particular gene. They use:

- **capital letters for dominant alleles**
- **lower case letters for recessive alleles**.

Example

For eye colour, brown is dominant and blue is recessive so B represents a brown allele and b represents a blue allele.

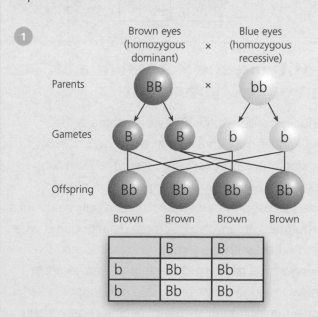

	B	B
b	Bb	Bb
b	Bb	Bb

Genetic Diagrams (cont)

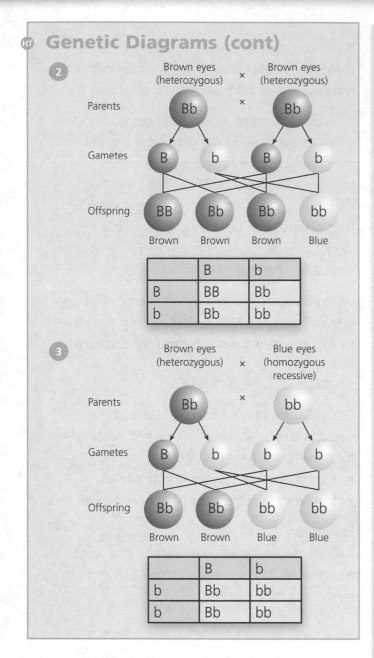

②

Brown eyes (heterozygous) × Brown eyes (heterozygous)

Parents: Bb × Bb

Gametes: B, b, B, b

Offspring: BB (Brown), Bb (Brown), Bb (Brown), bb (Blue)

	B	b
B	BB	Bb
b	Bb	bb

③

Brown eyes (heterozygous) × Blue eyes (homozygous recessive)

Parents: Bb × bb

Gametes: B, b, b, b

Offspring: Bb (Brown), Bb (Brown), bb (Blue), bb (Blue)

	B	b
b	Bb	bb
b	Bb	bb

Identifying the Recessive and Dominant Characteristics

It is usually possible to identify which characteristic is dominant and which is recessive in a genetic cross by looking at the offspring produced. The dominant characteristic will generally be expressed in greater numbers of offspring. However, this can only be useful if large numbers are involved. A more definite clue is when two individuals with the same characteristic produce offspring with a different characteristic – for example when two tongue-rollers produce a non tongue-roller. In this case the 'non tongue-rolling' gene is recessive and was not expressed in the parental features.

Inherited Diseases

Some disorders are caused by a 'faulty' gene, which means they can be **inherited** (i.e. passed from one generation of a family to the next), for example:

- **red–green colour blindness**, where the specialised cells in the eye cannot distinguish between red and green light
- **sickle cell anaemia**, which causes red blood cells to become sickle-shaped leading to circulatory problems and oxygen deficiency
- **cystic fibrosis**, which causes cell membranes to produce excess thick mucus which blocks airways. It also causes problems with digestion.

Cystic fibrosis is a disease that is due to a recessive characteristic. Two parents could appear to be completely healthy and unaffected by the disease, but if they both have a recessive allele for the disease, one of their children could inherit the disease from them. The parents would be known as **carriers**. (They do not have the disease themselves because their dominant 'normal' alleles protect them.) An affected child is likely to have a reduced quality of life and a reduced life expectancy. They will need life-long care and medication. However, modern drugs enable many affected people to lead happy and productive lives.

Conditions such as cystic fibrosis are mostly caused by **faulty alleles** that are **recessive**.

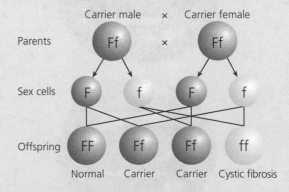

Carrier male × Carrier female

Parents: Ff × Ff

Sex cells: F, f, F, f

Offspring: FF (Normal), Ff (Carrier), Ff (Carrier), ff (Cystic fibrosis)

Knowing that there is a 1 in 4 chance that their child might have cystic fibrosis gives parents the opportunity to make decisions about whether to take the risk and have a child. However, this is a very difficult decision to make.

1 Jasmine accidentally picks up a hot tripod in her science lesson and immediately drops it.

a) Write down the name of this response. **[1]**

b) The responses in Jasmine's nervous system are shown in this sequence. Fill in the blank boxes with the missing structures. **[2]**

| Temperature receptor | → | | → | Relay neurone | → | Motor neurone | → | |

2 Adarsh is red–green colour blind. He finds it hard to distinguish between subtle shades of red and green. This condition is caused by a faulty gene on the X chromosome.

a) i) Adarsh takes a colour blindness test. The number '3' is drawn out in red on a background of crimson. Is he likely to be able to see the number? Give a reason for your answer. **[1]**

ii) Adarsh has a sister, Nateem. Nateem takes the same colour blindness test. Is she likely to be able to see the '3'? Use your knowledge of genetics to explain your answer. **[2]**

b) i) Like all humans, Adarsh has binocular vision. Describe how the arrangement of his pet rabbit's eyes is different. **[1]**

ii) The rabbit escapes from its hutch one day. Adarsh creeps up behind the animal to capture it. Explain why the rabbit sees Adarsh coming without needing to turn its head. **[1]**

HT **3** Study the graph below.

Smoking Habits and Lung Cancer Incidence in a European Country

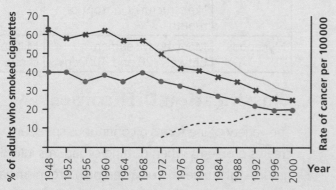

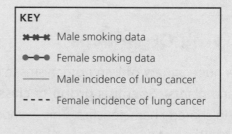

KEY
- ✖✖✖ Male smoking data
- ●●● Female smoking data
- —— Male incidence of lung cancer
- - - - Female incidence of lung cancer

a) Describe the patterns you can see in male and female smoking habits since 1948. **[2]**

b) Describe the differences in male and female cancer rates. **[2]**

c) Calculate the difference in the percentage of female smokers between 1948 and 1988. **[1]**

d) In what year did the country have the highest overall percentage of smokers? **[1]**

4 Two new drugs called 'RedU' and 'DDD' have been developed to help people lose weight. Clinical trials are carried out on DDD and RedU. The results are shown in the table.

Drug	Number of Volunteers in Trial	Average Weight Loss in 6 Weeks (kg)
RedU	3250	3.2
DDD	700	5.8
Placebo	2800	2.6

a) The scientific team concluded that DDD was a more effective weight loss drug. Do you agree? Use the data in the table to give a reason for your answer. **[2]**

b) The trial carried out was 'double blind'. Explain what this term means and why it is used. **[2]**

This module looks at:

- How organisms are classified according to shared characteristics.
- Energy flow through food chains, and pyramids of numbers and biomass.
- Cycling of nutrients in ecosystems, focusing on nitrogen and carbon.
- Interdependence – competition, predation, parasitism and mutualism.
- How organisms adapt to suit a changing environment.
- How variation leads to evolution, focusing on Darwin's theory of natural selection.
- How increasing human population is linked to pollution, and the problems of global warming, ozone depletion and acid rain.
- Sustainability and reasons why organisms are endangered or extinct.

Classifying Organisms

There is a huge variety of living organisms (e.g. animals and plants). Scientists **group** or **classify** them using shared characteristics. This is important because it helps us to:

- work out how organisms evolved on Earth
- understand how organisms coexist in ecological communities
- identify and monitor rare organisms that are at risk from extinction.

The Five-kingdom System

Observable characteristics are used to place organisms into the following **kingdoms**:

Kingdom	Features	Examples
Plants	Cellulose cell wall, use light energy to produce food	Flowering plants, trees, ferns, mosses
Animals	Multicellular, feed on other organisms	Vertebrates and invertebrates

Kingdom	Features	Examples
Fungi	Cell wall of chitin, produce spores	Toadstools, mushrooms, yeasts, moulds
Protoctista, protozoa	Mostly single celled organisms	Amoeba, Paramecium
Prokaryotes	No nucleus	Bacteria, blue–green algae

Arthropods

Arthropods are an important group of **invertebrates**. They include the following classes:

Class	Features	Examples
Insects	3 body parts, 2 pairs of wings, 3 pairs of legs	Locusts, bees, butterflies
Arachnids	2 body parts, 4 pairs of legs	Spiders, scorpions
Crustaceans	Segmented body divided into 3 parts, paired jointed limbs, 2 pairs of antenna-like structures in front of mouth	Crabs, lobsters, shrimps
Myriapods	Many body segments, between 20 and 400 legs	Millipedes, centipedes

Classification Difficulties

The variety of life forms a continuous spectrum which creates problems when putting organisms into definite groups. Difficulties occur when new organisms are discovered or fossils are found because observable features can sometimes be misleading. When the fossil **Archaeopteryx** was discovered, it was found to possess feathers like a bird but no beak, but it had claws like a reptile, so which group did it fit into?

The structure used to sort organisms into groups forms a hierarchy:

Kingdom   Largest group
Phylum
Class
Order
Family
Genus
Species   Smallest group

A useful way of understanding evolutionary relationships is to show them as an evolutionary tree:

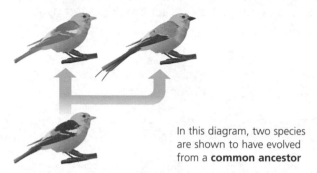

In this diagram, two species are shown to have evolved from a **common ancestor**

HT Evolutionary relationships can also be modelled by analysing multiple characteristics and feeding the information into sophisticated computer programs.

Overcoming Classification Difficulties

Systems of classification change over time due to new techniques and evidence. Some classification systems are natural and based on evolutionary systems. Others are artificial and used simply for the purposes of identification. To place organisms in their correct group, scientists look at evolutionary relationships; evidence can be found using sequences of bases in DNA and looking at the proteins that similar organisms possess. The more closely related they are, the more similar their DNA and proteins.

Other difficulties occur when dealing with **hybrids** and when trying to make a decision based on how far one species has evolved from another.

Species

Groups of organisms that share all the same characteristics are called **species**. Members of the same species are so similar that they can reproduce together. A species is a group of organisms that can freely interbreed to produce fertile offspring. Each species is given two Latin names. This method of naming species is called the **binomial** ('two name') system. For example, lions and tigers are closely related species so they share the Latin **genus** name *panthera*. All lions belong to the same species so they have the species

name *leo*. All tigers belong to the same species so they have the species name *tigris*.

New species are being identified all the time. Some ecosystems, such as the ocean depths, are relatively unexplored and potentially contain many undiscovered species.

Within a species there can be much (intra-specific) variation. This is a major factor in bringing about the evolution of new species. Members of the same species will have more features in common than they do with organisms of different species.

HT Some organisms from different species can mate and reproduce to give birth to a **hybrid**. However, hybrids are not fertile, i.e. they cannot successfully reproduce themselves and so cannot be called a new species. For example:

Male Donkey + Female Horse = Mule (a hybrid)

Bacteria as a group are not defined in the same way as other groups. This is because bacteria reproduce asexually and they do not need to produce sperm or eggs.

Species that share a lot of common features tend to live in similar habitats. They compete for resources and survive in the same conditions. However, closely related species can be found on different continents where conditions may be different so the species may have continually evolved to adapt to different conditions.

HT Organisms with similar characteristics are not necessarily descended from a common ancestor. They may have just evolved to survive in the same environment and so have developed similar structures. These organisms show an ecological connection. Whales, dolphins and sharks look quite similar, but they have descended from **different evolutionary ancestors**. Their similarities are due to sharing a similar environment.

Food Chains

Food chains show which organisms consume which other organisms. They also show the **transfer of energy** and materials from organism to organism. Energy from the Sun enters most food chains when green plants absorb sunlight to **photosynthesise**. Feeding passes this energy and **biomass** from one organism to the next along the food chain.

Grass   Rabbit   Stoat   Fox

The arrow shows the flow of energy and biomass along the food chain:

- All food chains start with a producer. This is usually a green plant but some producers can harness their energy for making food from chemical reactions, e.g. sulfur reactions which occur in undersea volcanoes.
- The rabbit is a herbivore (plant eater), the **primary consumer**.
- The stoat is a carnivore (meat eater), the **secondary consumer**.
- The fox is the top carnivore, the **tertiary consumer**.

Each consumer or producer occupies a **trophic level** (feeding level). However, sometimes organisms can occupy more than one trophic level. In the example above, we can see that a change in the numbers of one organism can affect the numbers of organisms in a connected trophic level.

Pyramid of Numbers

The number of organisms at each trophic level in the food chain can be shown as a **pyramid of numbers**. As we go up from one level to the next, the number of organisms decreases quite dramatically. This is because at each stage of the food chain energy is lost as heat (during respiration), via excretion (urine) and via egestion (removal of undigested food). These excretory parts, faeces and uneaten parts can act as the starting point for other food chains.

Pyramids of numbers usually look like ➊.

However, pyramids of numbers do not take into account the mass of the organisms, so it is possible to end up with some odd-looking inverted pyramids (see ➋).

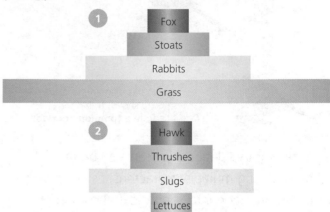

Because lots of slugs feed on one lettuce, the base of pyramid ➋ is smaller than the next stage. This happens because the lettuce is a large organism compared to the slug. This situation also happens when trees are at the start of a food chain.

Pyramid of Biomass

Pyramids of biomass deal with the dry mass of living material in the chain. They are always pyramid shaped because they consider the mass of the organisms. If the number of organisms is multiplied by their mass then the food pyramid ➋ above would give:

If enough information is given, the pyramid of biomass can be drawn to scale. You may be asked in an exam to interpret data on energy flow in food chains and construct pyramids of biomass from given data.

> HT In nature nothing is as clear-cut as these pyramids suggest. Organisms may occupy more than one trophic level, depending on which food chain in the web you are dealing with.

Consider this **food web**:

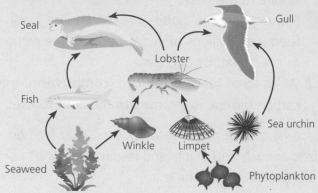

In this example, the seal is a secondary consumer for this food chain: Seaweed → Fish → Seal. However, the seal is a tertiary consumer for this food chain: Phytoplankton → Limpet → Lobster → Seal

When scientists construct pyramids of biomass they deal with **dry biomass**, because this represents the organic material produced from energy passed on from the trophic level beneath. A large proportion of an organism's mass is water, which can vary and does not reflect the energy passed on. Experiments to accurately measure dry mass involve heating the organism to a constant mass. This is obviously destructive and raises ethical issues.

Another complication is that excretory products and uneaten parts of organisms can be the starting points for other food chains, especially those involving detritivores and decomposers.

Efficiency of Energy Transfer

If you know how much energy is stored in the living organisms at each level of a food chain, the efficiency of energy transfer can be calculated by dividing the amount of energy used usefully (e.g. for growth) by the total amount of energy taken in:

$$\text{Energy efficiency (\%)} = \frac{\text{Energy used usefully}}{\text{Total energy taken in}} \times 100$$

Example
A sheep eats 100kJ of energy in the form of grass but only 9kJ becomes new body tissue; the rest is lost as faeces, urine or heat. Calculate the efficiency of energy transfer in the sheep:

$$\text{Energy efficiency} = \frac{9}{100} \times 100$$
$$= \mathbf{9\%}$$

Energy

The fox gets the last tiny bit of energy left after all the others have had a share. This explains why food chains rarely have fourth degree or fifth degree consumers – they would not get enough energy to survive.

The stoats run around, mate, excrete, keep warm, etc. and pass on about a tenth of all the energy they get from the rabbits.

The rabbits run around, mate, excrete, keep warm, etc. and pass on about a tenth of all the energy they get from the grass.

The Sun is the energy source for all organisms, but only a fraction of the Sun's energy is captured in photosynthesis.

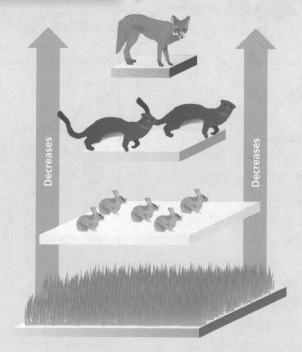

Biomass

The fox gets the remaining biomass.

The stoats lose quite a lot of biomass in faeces and urine.

The rabbits lose quite a lot of biomass in faeces and urine.

A lot of the biomass remains in the ground as the root system.

Recycling

In a stable community, the processes that remove materials are balanced by processes that return materials. So materials are constantly being **recycled**. For example, when animals and plants grow they take in elements from the soil into their bodies. When they die and decay, these elements are released and can be taken up by other living organisms to enable them to grow. This decay process is carried out by soil bacteria and fungi. This recycling of nutrients can take considerably longer in waterlogged or acidic soils because the conditions are not ideal for these microorganisms to survive. **Carbon** and **nitrogen** are two of these recycled elements.

The Carbon Cycle

The constant recycling of carbon is called the **carbon cycle**:

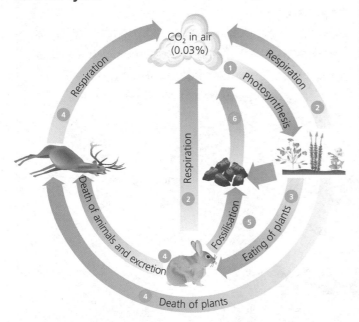

1. Carbon dioxide is removed from the atmosphere by green plants for photosynthesis.
2. Plants and animals respire, releasing carbon dioxide into the atmosphere.
3. Animals eat plants and other animals, which incorporates carbon into their bodies. In this way carbon is passed along food chains and webs.
4. Microorganisms feed on dead plants and animals, causing them to decay and release carbon dioxide into the air. (The microorganisms respire as they feed.)
5. Some organisms' bodies are turned into fossil fuels over millions of years, trapping the carbon as coal, peat, oil and gas.
6. When fossil fuels are burned (**combustion**) the carbon dioxide is returned to the atmosphere.

Bacteria and fungi are decomposers. They feed on dead animals and plants and respire, releasing carbon dioxide into the air.

HT Carbon is also recycled in the sea:

1. Marine organism shells are made of carbonates. They drop to the sea bed as the organisms die.
2. The shells fossilise into limestone rock.
3. Volcanic eruptions heat the limestone and release carbon dioxide into the atmosphere.
4. Acid rain chemically weathers limestone buildings and rocks, releasing carbon dioxide.
5. Oceans can absorb carbon dioxide directly from the air, acting as **carbon sinks**.

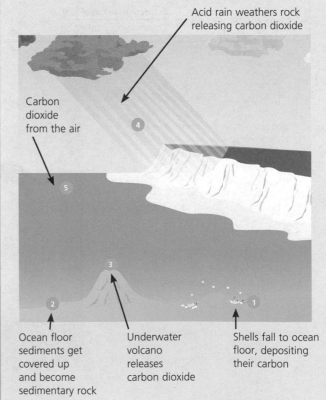

Acid rain weathers rock releasing carbon dioxide

Carbon dioxide from the air

Ocean floor sediments get covered up and become sedimentary rock

Underwater volcano releases carbon dioxide

Shells fall to ocean floor, depositing their carbon

The Nitrogen Cycle

The air contains approximately 78% nitrogen. Nitrogen is a vital element of all living things and is used in the production of proteins, which are needed for growth in plants and animals. The **nitrogen cycle** (see diagram opposite) shows how nitrogen and its compounds are recycled in nature:

1. Plants absorb nitrates from the soil to make protein for growth.
2. Animals eat plants and use the nitrogen to make animal protein. In this way, nitrogen is passed along food chains and webs.
3. Dead animals and plants are broken down by decomposers, releasing nitrates back into the soil.

There is a lot of nitrogen stored in the air, but animals and plants cannot use it because it is so unreactive.

The Role of Bacteria in the Cycle

Nitrogen-fixing bacteria convert atmospheric nitrogen into nitrates in the soil. Some of these bacteria live free in the soil while some are associated with the root systems of certain plants.

Decomposers convert urea and proteins into ammonia. **Nitrifying bacteria** then convert the ammonia to ammonium compounds and, subsequently, nitrates in the soil. **Denitrifying bacteria** convert nitrates into atmospheric nitrogen, and ammonium compounds into atmospheric nitrogen.

Root Nodules

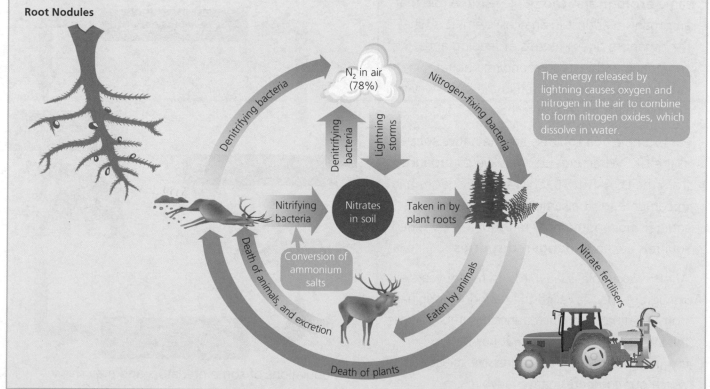

B2 | Interdependence

Factors Affecting Population Size

The size of any population of plants or animals will change over time. It can be affected by how well the populations compete for resources. For animals these resources include food, shelter and mates. For plants they include water, minerals and light.

Competitors

Similar organisms living in the same habitat will be in close **competition**. Even organisms within the same species may compete in order to survive and breed.

You may be asked to interpret data showing how animals and plants can be affected by competition.

> **HT** Competition can be described as **intra-specific** (between members of the same species) or **inter-specific** (between members of different species). Intra-specific competition can often be the most significant type, because members of the same species have the same needs and therefore there is competition for exactly the same niche.
>
> Similar organisms living in the same habitat with the same prey and nesting sites occupy the same **ecological niche**. A niche is the role that an organism adopts in an ecosystem. It is what the organism does in terms of feeding and competing within its surrounding habitat. Organisms that are adapted in the same way will occupy similar ecological niches.
>
> For example, red squirrels are the native species in the UK. When grey squirrels were introduced from the USA in 1876, both squirrel species had to compete for the same resources. Grey squirrels now outnumber red squirrels 66:1. Red squirrels are an **endangered species**.
>
> Another example is ladybirds. In recent years an immigrant species called the harlequin (originally from Asia) has been increasing in number across the UK's south coast, where it has begun forming swarms. It is an aggressive species, which out-competes native ladybirds.

Predators and Prey

Animals that kill and eat other animals are called **predators** (e.g. foxes, lynx), while the animals that are eaten are called **prey** (e.g. rabbits, snowshoe hares).

Many animals can be both predator and prey, e.g. a stoat is a predator when it hunts rabbits and it is the prey when it is hunted by a fox.

Predator – Stoat   **Predator** – Fox

Prey – Rabbit   **Prey** – Stoat

In nature there is a delicate balance between the population of a predator (e.g. lynx) and its prey (e.g. snowshoe hare). However, the prey will always outnumber the predators.

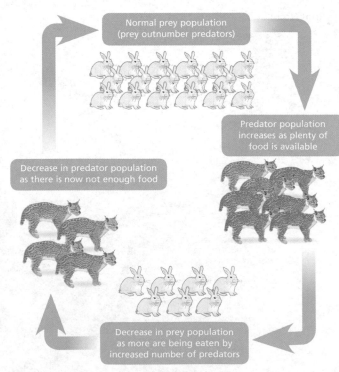

Normal prey population (prey outnumber predators)

Predator population increases as plenty of food is available

Decrease in predator population as there is now not enough food

Decrease in prey population as more are being eaten by increased number of predators

Populations of some predators and prey show cyclical changes in numbers.

Predator–Prey Cycles

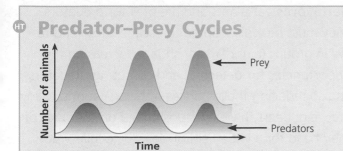

The number of predators and prey follow a classic population cycle. When there are lots of hares, the lynx have more food so numbers increase. As they eat lots of hares, the hare numbers then decrease. There will always be more hares than lynx and the population peak for the lynx will always come after the population peak for the hare – because it is cause and effect, they will always be out of phase.

Interdependence

The interdependence of organisms determines their **distribution** (how spread out they are), and their **abundance** (how many of them there are).

Below is an example of a food pyramid. An increase or decrease in the number of plants or animals at one stage in the food chain can affect the rest of the food chain. For example, if the rabbits were killed off by a disease:

* the number of stoats would decrease because they would lose one of their food sources; this would then affect the foxes
* the stoats would have to find another source of food
* the number of grass plants would increase because fewer rabbits would be eating them.

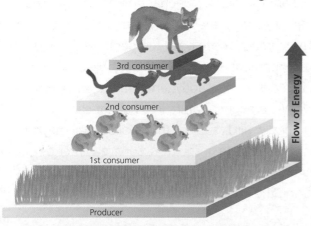

Parasitic Relationships

Parasites are organisms that live off another organism (the **host**). This can sometimes make the host organism ill or even kill it. For example, humans (the host) can contract tapeworms (the parasite) by eating pork infected with tapeworm larvae (bladderworms). Once inside a person's body, the larvae attach to the wall of the gut, and a young tapeworm grows, which absorbs food from the person's gut.

Fleas are **ectoparasites**. They live on the skin of animals such as dogs and cats, feeding on the host's blood and causing the host to scratch in an attempt to remove the fleas.

Mutualistic Relationships

In **mutualistic relationships**, two organisms form a relationship from which both benefit. For example, oxpecker birds get a ready supply of food from ticks and flies on a buffalo's skin. The oxpecker birds are known as a **cleaner species**. The buffalo benefits because the birds get rid of the pests and provide an early-warning system by hissing when lions or other predators approach.

Nitrogen-fixing bacteria in the root nodules of leguminous plants are another example. The bacteria gain sugars and the plants gain compounds containing nitrogen.

Insects such as bees have a mutualistic relationship with flowering plants: the flower has a means of pollination, while the insect gets food in the form of nectar.

The interdependence of organisms dictates where they will be found and in what numbers. Leguminous plants, e.g. the pea plant, have root nodules that contain nitrogen-fixing bacteria, which take sugars from the plant to use in respiration and convert nitrogen into nitrates that the plant can absorb. Leguminous plants are able to survive in nitrogen-poor soils because of their mutualistic relationship with the bacteria.

Adaptations

Adaptations are special features or behaviour that make an organism particularly well suited to its environment, and better able to compete with other organisms for limited resources. An adaptation can be thought of as a **biological solution** to an environmental challenge. Evolution provides the solution and makes species fit their environment.

Animals have developed in many different ways to become well adapted to their environment and help them to survive. Look at the polar bear and its life in a very cold climate:

- Small ears and large bulk both reduce the surface area to volume ratio to reduce heat loss.
- Large amount of insulating fat (blubber).
- Thick white fur for insulation and camouflage.
- Large feet to spread its weight on snow and ice.
- Fur on the soles of its paws gives insulation and grip.
- Powerful legs so it is a good swimmer and runner, which enables it to catch its food.
- Sharp claws and teeth to capture prey.

As plants and animals become better adapted to their environment their abundance and distribution can increase.

Predators and Prey

Carnivores eat other animals, so they must be successfully adapted to be good predators. For example, lions are excellent predators. They are built for bursts of speed but are camouflaged. They have sharp teeth and claws to grab and kill their prey. The lion's eyes are positioned at the front of its head, providing three-dimensional vision and accurate perception of size and distance. Predators use hunting strategies, e.g. wolves hunt in packs so they can catch and kill large prey. Some predators have specialised breeding strategies, e.g. most owls time their breeding so that their young are able to hunt at the time of their prey's breeding season.

Herbivores eat plants and are prey for predators. They must be well adapted to escape. Prey tend to live in groups (e.g. herds or shoals), increasing the opportunities for detecting and confusing predators and so reducing their chances of being caught. Many prey, e.g. deers, have eyes positioned on the sides of their head for a wide field of view (binocular vision) and are built for speed so they can escape.

Some prey, e.g. bees and scorpions, use defence mechanisms such as stings or poisons to defend themselves against attack. Such organisms may have **cryptic colouration** which camouflages them from predators. Some organisms, e.g. the hoverfly, have warning colouration even though they are harmless. This survival strategy is called **mimicry**.

Prey tend to have more young than predators as many get eaten. Other, more unusual, adaptations extend to an organism's life cycles. Some bird species, such as the starling, can reproduce in a very short time to take advantage of favourable conditions; they all produce offspring at the same time, a process called **synchronous breeding**. Giving birth to live young rather than laying eggs is also advantageous: the offspring's development is more advanced when they are exposed to the environment, so they can fend for themselves earlier rather than be dependent on parents.

Adaptations' Environmental Stresses

Some organisms are adapted to cope with high temperature by having a large surface area to volume ratio. For example the elephant has large thin ears to aid heat loss from the blood vessels that run through them. Animals whose body temperature depends on the environment, e.g. the crocodile, will bask in the sun in order to raise body temperature and become active. Where temperature in a habitat changes dramatically, birds such as the swallow will **migrate**. Many mammals, e.g. squirrels, survive winter by lowering their body temperature and **hibernating**, reducing their dependence on scarce food supplies.

HT Further adaptations to climate include:

- Penguins operate complex **counter-current heat exchange systems** to minimise heat loss. Warm blood entering the feet flows past cold blood leaving the feet, and warms it up. The warmed-up blood re-enters the body and doesn't affect the penguin's core temperature.
- Biochemical adaptations including **extremophiles**, which are organisms that can survive extreme environmental conditions, e.g. organisms that have optimum temperatures for enzymes that are lower or higher than 37°C.
- Some organisms, e.g. icefish, have **antifreeze chemicals** in their bodies, which lower the freezing point of body fluids.

The size of an organism can have a strong bearing on its ability to survive in extreme temperatures. **Large** animals have a **low** surface area-to-volume ratio – putting it simply, there is a proportionately greater amount of body tissue in a larger organism. This makes it easier for them to retain heat in their bodies, but harder to absorb it in the first place. This has implications for large 'cold-blooded' animals (e.g. reptiles) because they rely on heat from the environment to raise their body temperature. **Small** animals, on the other hand, have a **large** surface area-to-volume ratio – they lose heat rapidly to the environment, but can gain it quickly. For example, a shrew living in a cold climate will lose heat rapidly, but will fare better than a small lizard, which cannot control its core body temperature internally.

Examples of Desert Adaptations

The Cactus

- Rounded shape reduces water loss by giving a small surface area to volume ratio.
- Thick waxy cuticle reduces water loss.
- Stores water in a spongy layer inside its stem to resist drought.
- Leaves reduced to spines to reduce water loss and to protect the cactus from predators.

- Green stem so that the plant can photosynthesise without leaves.
- Long roots to reach water.

The Camel

- Large surface area to volume ratio to increase heat loss.
- Body fat stored in a hump so there is very little in the insulating layer beneath the skin.
- Loses very little water in sweat or urine.
- Able to tolerate changes in body temperature so it does not need to sweat so much.
- Large feet to spread its weight on the sand.
- Bushy eyelashes and hair-lined nostrils to stop sand from entering.

Desert Lizards and Desert Rats

Some desert lizards have evolved behavioural methods of reducing heat gain. For example, many are nocturnal or come out at dusk or dawn, spending the day under the sand. Others alternately lift their legs so that only two feet are in contact with the hot sand at any one time.

To conserve water, desert rats have evolved methods of extracting the water released from respiration when seeds and roots are consumed. They also have specialised kidneys which re-absorb most of the water in their urine. Some insects position themselves with their abdomens raised at night so that condensed water that has accumulated on their bodies runs into their mouths.

HT Specialists and Generalists

Some organisms have evolved as **specialists** that are well suited to only certain habitats and niches. Should the niche disappear over time then such an organism is at a disadvantage and may become endangered or extinct. The classic example is that of the panda, which has a very specialised diet of bamboo.

A different strategy is adopted by the horseshoe crab which has survived many millions of years due to its **generalist** nature. It has a wide variety of food types and can migrate to different parts of the ocean during its life cycle.

The Theory of Evolution

The **theory of evolution** states that all living things that exist today, and many more that are now extinct, evolved from simple life forms, which first developed more than 3.5 billion years ago.

Evolution is the slow, continual change of organism groups over a very long period to become better **adapted to their environment**.

Within a population of organisms there is a range of variations of individuals, which is caused by genes. Some differences will be beneficial; some will not. Beneficial characteristics make an organism more likely to survive and pass on their genes to the next generation. This is especially true if the environment is changing. This ability to survive is called **survival of the fittest**. Species that are not well adapted to their environment may become extinct. This process of change is summed up in a theory called **natural selection**, put forward by **Charles Darwin** in the 19th century. Many theories have tried to explain how life might have come about in its present form. However, Darwin's theory is accepted by most scientists today. This is because it explains a wide range of observations and has been discussed and tested by many scientists.

At the time, the reaction to Darwin's theory, particularly from religious authorities, was hostile because they felt he was saying that 'men were descended from monkeys' (although he was not) and that he was denying God's role in the creation of man. This meant that his theory was only slowly and reluctantly accepted by many people in spite of his many eminent supporters.

Darwin's theory can be reduced to five ideas illustrated by the examples that follow.

Peppered Moths
* **Variation** – Most peppered moths are pale and speckled, and are easily camouflaged amongst the lichens on silver birch tree bark. There are some rare, dark-coloured varieties (which originally arose from genetic mutation) but they are easily seen and eaten by birds.

* **Competition** – In areas with high levels of air pollution, lichens die and the bark becomes discoloured by soot. The lighter versions are now put at a competitive disadvantage.
* **Survival of the fittest** – The dark (melanic) moths are now more likely to avoid detection by predators.
* **Inheritance** – The genes for dark colour are then passed on to offspring and gradually become more common in the general population.
* **Extinction** – If the environment remains polluted then the lighter form is more likely to become extinct.

Warfarin-resistant Rats
Rats are a pest in urban and rural environments. One way to deal with them is to poison them. Warfarin is an effective poison that prevents blood clotting so the rats slowly bleed to death. Warfarin has been so widely used that a breed of 'super rats' has emerged.

* **Variation** – Mutant rats, able to resist warfarin, arise in the population.
* **Competition** – Mutant rats are not killed by the poison.
* **Survival of the fittest** – Warfarin-resistant rats survive, but the non-resistant breed do not.
* **Inheritance** – Resistant rats pass on their warfarin-resistant genes.
* **Extinction** – Non-resistant rats may eventually become extinct. There are already large populations of warfarin-resistant rats in the UK.

Penicillin-resistant Bacteria

The resistance of some bacteria to antibiotics is an increasing problem. MRSA bacteria have become more common in hospital wards and are difficult to eradicate.

- **Variation** – Bacteria mutate by chance, giving them a resistance to antibiotics.
- **Competition** – The non-resistant bacteria are more likely to be killed by the antibiotic.
- **Survival of the fittest** – The antibiotic-resistant bacteria survive and reproduce more often.
- **Inheritance** – Resistant bacteria pass on their genes to a new generation. The gene becomes more common in the general population.
- **Extinction** – Non-resistant bacteria are replaced by the newer, resistant strain.

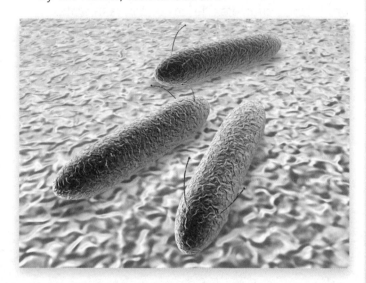

New Species Emerge

Species become more and more specialised as they evolve, adapting to their environmental conditions.

Groups of the same species that are separated from each other by physical boundaries like mountains or seas will not be able to breed and share their genes. This is called **geographical isolation**. Over long periods of time, the separate groups may specialise so much that they cannot successfully breed any longer and so two new species have formed – this is **reproductive isolation**.

Jean-Baptiste Lamarck

Before Darwin developed his theory, **Lamarck** suggested that evolution happened by the inheritance of acquired characteristics:

- Organisms change during their lifetime as they struggle to survive.
- These changes are passed on to their offspring.

For example, he said that when giraffes stretch their necks to reach leaves higher up, this extra neck length is passed on to their offspring.

Lamarck's theory was rejected because there was no evidence that the changes that happened in an individual's lifetime could alter their genes and so be passed on to their offspring.

Charles Darwin

Darwin made four very important observations:

1. Many living things produce far more offspring than actually survive to adulthood.
2. In spite of this, population sizes remain fairly constant due to predation, etc.
3. There is variation between members of the same species.
4. Characteristics can be passed on from one generation to the next.

From these observations, Darwin deduced that all organisms were involved in a struggle for survival in which only the best-adapted organisms would survive, reproduce and pass on their characteristics. This formed the basis for his famous theory.

The theory of natural selection has developed further since Darwin's time as new discoveries have been made. The study of inheritance and knowledge about DNA has added significantly to our understanding of this basic process.

The Population Explosion

The human population is increasing exponentially (i.e. at a rapidly increasing rate). This is because birth rates exceed death rates by a large margin. This is creating three major issues:

1. The use of finite resources like fossil fuels and minerals is increasing.
2. The production of pollution, in particular household waste, sewage, sulfur dioxide and carbon dioxide, is increasing.
3. Ever-increasing competition for basic resources, i.e. food and water.

HT The exponential growth of the human population means that increase in demand for resources is also exponential:

- Raw materials are being used up increasingly quickly.
- Pollution and waste are building up at an alarming rate.
- As resources become in short supply, increased competition makes things like food and water more expensive.

Although the developed countries of the world (e.g. USA, UK, France and Japan) have a small proportion of the world's population, they use the greatest amount of resources and produce the largest proportion of pollution. Many countries are now trying to agree limits on the production of pollution to prevent it becoming worse in the future.

Ozone Depletion

Ozone is a natural gas, found high up in the Earth's atmosphere, that prevents too many harmful ultraviolet (UV) rays reaching the Earth.

In the 1980s, scientists noticed that the ozone layer was getting thinner and that skin cancer cases were increasing. This is because the thinner ozone layer is unable to block out as many UV rays.

Many people blame the use of **CFCs** (chlorofluorocarbons) in factories, fridges and aerosol cans for this change in the ozone layer. Much has been done since the 1980s to reduce the use of CFCs, but ozone depletion is still a problem – particularly over the Arctic and Antarctic – and it will take a long time for the ozone layer to recover.

Acid Rain

When coal or oils are burned, sulfur dioxide is produced. Sulfur dioxide gas dissolves in water to produce **acid rain**. Acid rain can:

- damage trees, stonework and metals
- make rivers and lakes acidic, which means some organisms can no longer survive.

The acids can be carried a long way away from the factories where they are produced. Acid rain falling in one country could be the result of fossil fuels being burned in another country.

The Greenhouse Effect and Global Warming

Ultraviolet light from the Sun reaches Earth in the form of radiation. Some of this energy is reflected back out towards space as infrared radiation. When it reaches the atmosphere some rays pass through, but others are trapped in. It is these trapped rays that keep the Earth warmer than it would be otherwise. This is known as the **greenhouse effect**.

However, because the amount of carbon dioxide and other gases (such as methane) in the atmosphere has now increased, this has led to more of the energy being reflected back. This is known as **global warming** because the Earth's average temperature is increasing. A temperature rise of only a few degrees Celsius may lead to climate change and a rise in sea levels.

Global Warming

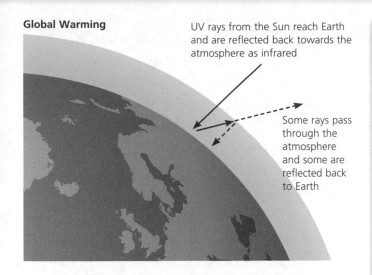

UV rays from the Sun reach Earth and are reflected back towards the atmosphere as infrared

Some rays pass through the atmosphere and some are reflected back to Earth

> In recent times the term **carbon footprint** has been used to describe the amount of greenhouse gases given off by individuals, organisations and industry in a certain amount of time.

Living Organisms as Indicators

Pollution reduces the number and type of organisms that can survive in a particular place. Some organisms are very sensitive to pollution, so they die. Other species have evolved to resist the toxic effects of pollution and can survive. These are called **indicator species**. Two examples of indicator species are lichens and freshwater insect larvae.

Lichens

Different types of lichens have differing levels of sensitivity to sulfur dioxide pollution in the air. The more resilient varieties are able to survive even when high levels of sulfur dioxide are found in the air. The more sensitive ones cannot, which causes them to die.

The distribution of lichens acts as a good indicator of the concentration of sulfur dioxide in the air. For example, if you look at the lichens found on trees close to a busy road, you will find just a few varieties of lichen. However, as you get further away from the road, you will notice that you will find more, and a wider variety of, lichens on the trees. This is because the sulfur dioxide is less intense further away from the road, so more varieties are able to flourish.

Insect Larvae

Insect larvae can act as an indicator of water pollution. When sewage works outflow into streams, this pollutes the water by altering the levels of nitrogen compounds in the stream and reducing oxygen levels. This then has an impact on the organisms that can survive in the stream.

Organisms that can cope with pollution include the rat-tailed maggot, the bloodworm, the waterlouse and the sludgeworm. However, some organisms are very sensitive to this type of water pollution so they are not found in areas where the levels are high. Organisms such as mayfly and stonefly larvae are killed by high levels of water pollution (they cannot tolerate low oxygen levels), so they are indicators of clean water.

Measuring Pollutant Levels

Scientists can measure the concentrations of pollutants by direct chemical tests. For example, water can be tested for pH and samples can be assessed for metal ion content, e.g. mercury.

In addition, the occurrence of certain indicator species can be observed by sampling an area and noting whether the species is present or not, together with the overall numbers of the species.

> The advantage of direct testing is that it produces an accurate measurement of pollution that can be compared with values from reference sources. The disadvantage is that many techniques require samples being taken to the lab and lengthy procedures being carried out.
>
> Using biological indicators gives a rapid assessment of pollution levels in the field as observations and counts can be carried out on the spot. But the data might not be conclusive on its own, because there are other factors that might determine the presence or absence of an indicator species, e.g. competition, predation.
>
> In an examination you may be asked to interpret data on indicator species.

Sustainable Development

As the world's population grows there is an increasing demand for food and energy, and more waste products are created.

A **sustainable resource** is one that can be used and replaced, ensuring that it is not used up completely.

Sustainable development is concerned with ensuring that resources can be maintained in the long term at a level that allows appropriate consumption or use by an increasing population. This often requires limiting exploitation by using quotas or ensuring the resources are replenished or restocked. Therefore, harm to the environment is minimised.

Cod in the North Sea

The UK has one of the largest sea fishing industries in Europe. To ensure that the industry can continue and that cod stocks can be conserved, quotas are set to prevent over-fishing. In 2006 the European Union Fisheries Council made changes including:

- increasing mesh size to prevent young cod being caught before they reach breeding age
- increasing quotas of types of fish other than cod.

Pine Forests in Scandinavia

Scandinavia uses a lot of pine wood to make furniture and paper and to provide energy. To ensure the long-term economic viability of pine-related industries, companies replenish and restock the pine forests by planting a new sapling for each mature tree they cut down.

How effective sustainable development is can be influenced by many factors – such as planning initiatives and cooperation at local, national and international levels. This has great implications for the protection of endangered species, e.g. quotas can be set for whaling.

Exponential increase in human population size makes sustainable development quite a challenge. As the population increases rapidly, the demand for food and energy also increases. The quickest and cheapest ways to meet these demands aren't always the most sustainable.

Protecting Endangered Species

Species are at risk of **extinction** if the number of individuals or habitats falls below a critical level.

Reasons for a decline in species numbers include:
- Pollution, e.g pollutants can accumulate in marine mammals.
- Over-hunting by humans
- Destruction of habitats, e.g by logging companies.
- Climate change
- Increased competition for food, shelter etc.

Endangered species can be protected from extinction in the following ways:
- **Education** helps to protect endangered species by promoting sustainable development.
- Animals can be bred in **captivity** and returned to their natural habitat to breed.
- Protecting the **natural habitat** (for example, by creating Sites of Special Scientific Interest) or creating artificial **ecosystems** (e.g. zoos, aquariums) to provide good conditions for the species to live in.
- **Hunting** legally protected species is **prohibited** and enforced by patrolling wardens in some cases.
- **Seedbanks** are created where vast numbers of plant seeds are stored in ideal conditions, e.g. Svalbard, Norway has a seed vault.

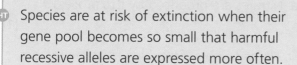

Species are at risk of extinction when their gene pool becomes so small that harmful recessive alleles are expressed more often.

In an exam you may be asked to evaluate a particular conservation programme in terms of:
• genetic variation of key species
• viability of populations (how likely they are to survive)
• available habitats
• interaction within species.

Whales

There are many different whale species, but some of these are now endangered i.e. they are close to extinction. The main causes of whale deaths include:
• getting entangled in fishing nets and drowning
• being affected by pollutants in the sea – whales are at the top of the food chain, so they accumulate pollutants from their food
• colliding with ships during migration
• effects of climate change affecting food sources
• culling and hunting – to reduce the population size, thus preventing competition with the fishing industry
• hunting – to provide food.

Live whales can be a big tourist attraction, but dead whales can be used for food, oil and to make cosmetics. Conservation campaigns have made people much more aware of the plight of whales.

One method of conserving whales has been to keep them in captivity and some zoos have had success with captive breeding. It enables whale behaviour to be studied so we can understand them and protect them more efficiently. However, captive whales suffer a huge loss of freedom and do not behave naturally; many are tamed and trained to perform for the public.

In an exam you may be asked to interpret data linking whale distribution and feeding habits.

Conservation Programmes

The needs of a growing population and the resulting demands of energy, food and waste products are often in conflict with the idea of **sustainable development**. Sustainability requires planning and cooperation at local, national and international levels. Conservation:
• protects the human food supply by maintaining the genetic variety of crops, animals and plants
• stabilises ecosystems by ensuring minimal damage to food chains and habitats (in this way sustainable development helps to protect species too)
• studies plants that might be useful to develop treatments for diseases
• protects the culture of indigenous people living in threatened habitats such as the Amazonian rainforest.

Conservation is difficult. For example, we need to know much more about whales to protect them effectively. At the moment, our knowledge of how whales communicate over large distances, how they migrate and how they dive and survive at extreme depths (whales are air-breathing mammals) is quite limited. Studying whales in captivity can be misleading but is easier than trying to study whales living freely in the ocean!

Different countries take very different points of view on issues like whaling. The International Whaling Commission makes laws to protect whale species and sets quotas for hunting. It is very difficult to enforce these laws though because it is impossible to police the world's oceans. It is also difficult to get all countries to agree.

Many countries agree that whale hunting is unnecessary. But, some countries like Iceland, Norway and Japan disagree with a ban on killing whales. They feel it is necessary to preserve the fishing industry and carry out 'research culls' to investigate the effect of whale population size on fish stocks.

1 a) When animals die, their bodies are decomposed by bacteria. Nitrates are produced by this process. Describe how plants use this nitrate. **[1]**

b) Decomposition also releases carbon dioxide back into the air. Name the process which transfers the gas into the bodies of plants. **[1]**

2 An environmental group is carrying out a survey of air quality close to a coal-fired power station. The group know that different lichens are sensitive to different amounts of pollution as follows: *fruticose* – very sensitive; *foliose* – sensitive; *crustose* – not very sensitive.

The extent of the presence of the three different lichens is assessed by looking at the trees surrounding the power station. The data collected by the group is shown in the table below:

Tree species	Number of trees sampled	Number of trees with fruticose lichen	Number of trees with foliose lichen	Number of trees with crustose lichen	Number of trees covered with any lichen	% of trees covered with any lichen
Ash	27	2	9	15	10	37%
Sycamore	15	1	4	10	13	
Pine	52	0	0	0	0	0%

a) Complete the table by working out the percentage of sycamore trees covered with any lichen. **[1]**

b) The group seemed to think that the number of fruticose lichens showed that there was a lot of air pollution. Do you agree? Give a reason for your answer. **[1]**

c) Describe two ways in which the group could obtain more reliable data. **[2]**

3 Scientists believe that the whale may have evolved from a horse-like ancestor which lived in swampy regions millions of years ago. Suggest how whales could have evolved from a horse-like mammal. In your answer, use Darwin's theory of natural selection. **[4]**

4 Global warming could lead to severe melting of polar ice caps. Explain how this might affect polar bears and why. **[6]**

✎ *The quality of written communication will be assessed in your answer to this question.*

5 An environmental scientist observes and measures a kingfisher and fish population in a county's rivers over 10 years. She records her results and processes them as a graph.

a) How many fish were recorded in the third year? **[1]**

b) Describe how the size of the kingfisher population affects the size of the fish population. **[1]**

HT c) The scientist took her measurements by ringing and observing kingfishers on three rivers in the county. Fish numbers were estimated by counting the different species which anglers landed along the banks of the three rivers. Describe the limitations of these methods and suggest two ways in which the methods could be improved. **[6]**

✎ *The quality of written communication will be assessed in your answer to this question.*

You need to have a good understanding of the concepts (ideas) on the next four pages, so make sure you revise this section before each exam.

Elements and Compounds

An **element** is a substance made up of just one type of **atom**. Each element is represented by a different chemical symbol, for example:

- Fe represents iron
- Na represents sodium.

Atoms have a positive nucleus orbited by negative electrons.

These elements (and their chemical symbols) are all arranged in the **Periodic Table** (see below).

Compounds are substances formed from the atoms of two or more elements, which have been joined together by one or more chemical bonds, for example, H_2O, $CaCO_3$ and $C_6H_{12}O_6$.

Ions are atoms or small molecules that have a charge, for example, Na^+, Cl^-, NH_4^+ and SO_4^{2-}.

A positive ion is formed when an atom loses electrons; a negative ion is formed when an atom gains electrons.

Covalent bonds are formed when two atoms share a pair of electrons. (The atoms in molecules are held together by covalent bonds.)

Ionic bonds are formed when atoms lose or gain electrons to become charged ions; the positive ions attract the negative ions.

The Periodic Table

1	2											3	4	5	6	7	0
																	4 **He** helium 2
7 **Li** lithium 3	9 **Be** beryllium 4											11 **B** boron 5	12 **C** carbon 6	14 **N** nitrogen 7	16 **O** oxygen 8	19 **F** fluorine 9	20 **Ne** neon 10
23 **Na** sodium 11	24 **Mg** magnesium 12											27 **Al** aluminium 13	28 **Si** silicon 14	31 **P** phosphorus 15	32 **S** sulfur 16	35.5 **Cl** chlorine 17	40 **Ar** argon 18
39 **K** potassium 19	40 **Ca** calcium 20	45 **Sc** scandium 21	48 **Ti** titanium 22	51 **V** vanadium 23	52 **Cr** chromium 24	55 **Mn** manganese 25	56 **Fe** iron 26	59 **Co** cobalt 27	59 **Ni** nickel 28	63.5 **Cu** copper 29	65 **Zn** zinc 30	70 **Ga** gallium 31	73 **Ge** germanium 32	75 **As** arsenic 33	79 **Se** selenium 34	80 **Br** bromine 35	84 **Kr** krypton 36
85 **Rb** rubidium 37	88 **Sr** strontium 38	89 **Y** yttrium 39	91 **Zr** zirconium 40	93 **Nb** niobium 41	96 **Mo** molybdenum 42	[98] **Tc** technetium 43	101 **Ru** ruthenium 44	103 **Rh** rhodium 45	106 **Pd** palladium 46	108 **Ag** silver 47	112 **Cd** cadmium 48	115 **In** indium 49	119 **Sn** tin 50	122 **Sb** antimony 51	128 **Te** tellurium 52	127 **I** iodine 53	131 **Xe** xenon 54
133 **Cs** caesium 55	137 **Ba** barium 56	139 **La*** lanthanum 57	178 **Hf** hafnium 72	181 **Ta** tantalum 73	184 **W** tungsten 74	186 **Re** rhenium 75	190 **Os** osmium 76	192 **Ir** iridium 77	195 **Pt** platinum 78	197 **Au** gold 79	201 **Hg** mercury 80	204 **Tl** thallium 81	207 **Pb** lead 82	209 **Bi** bismuth 83	[209] **Po** polonium 84	[210] **At** astatine 85	[222] **Rn** radon 86
[223] **Fr** francium 87	[226] **Ra** radium 88	[227] **Ac*** actinium 89	[261] **Rf** rutherfordium 104	[262] **Db** dubnium 105	[266] **Sg** seaborgium 88	[264] **Bh** bohrium 107	[277] **Hs** hassium 108	[268] **Mt** meitnerium 109	[271] **Ds** darmstadtium 110	[272] **Rg** roentgenium 111							

1 **H** hydrogen 1

Fundamental Chemical Concepts

Formulae

Chemical symbols are used with numbers to write **formulae** that represent the composition of compounds. Formulae are used to show:

- the different elements in a compound
- the number of atoms of each element in the compound
- the total number of atoms in the compound.

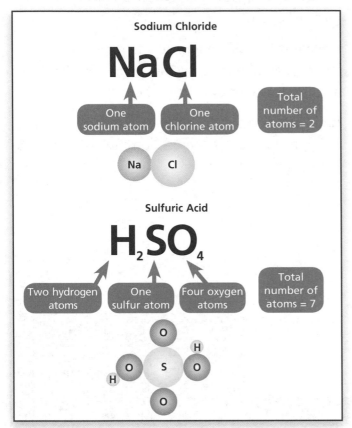

If there are brackets around part of the formula, everything inside the brackets is multiplied by the number outside the bracket.

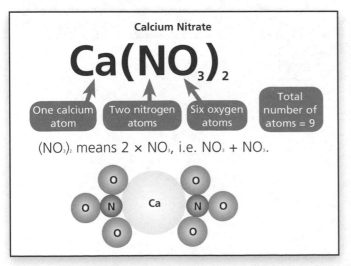

Displayed Formulae

A **displayed formula** is another way to show the composition of a molecule.

A displayed formula shows:

- the different types of atom in the molecule, e.g. carbon, hydrogen
- the number of each different type of atom
- the covalent bonds between the atoms.

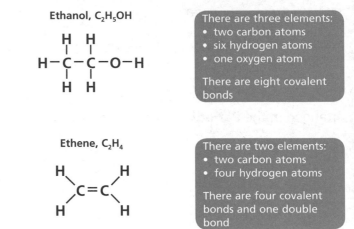

Equations

In a chemical reaction, the substances that you start with are called **reactants**. During the reaction, the atoms in the reactants are rearranged to form new substances called **products**.

Chemists use equations to show what has happened during a chemical reaction. The reactants are on the left side of the equation, and the products are on the right.

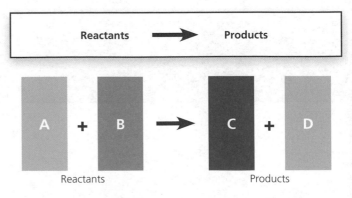

No atoms are lost or gained during a chemical reaction so equations must be **balanced**: there must always be the same number of atoms of each element on both sides of the equation.

Fundamental Chemical Concepts

Writing Balanced Equations

Example

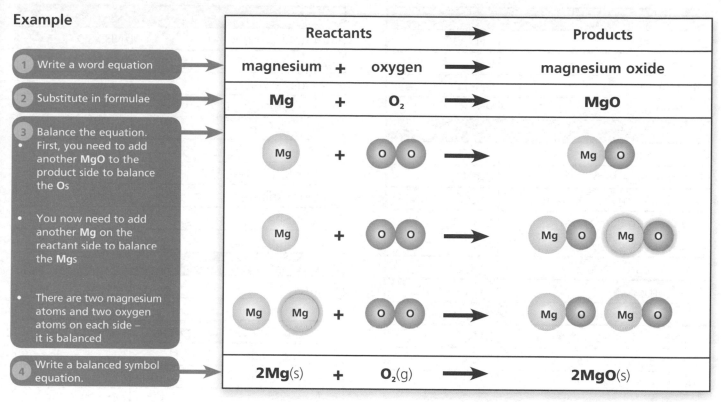

	Reactants		→		Products
1 Write a word equation	magnesium	+ oxygen	→		magnesium oxide
2 Substitute in formulae	**Mg**	+ **O₂**	→		**MgO**

$$2Mg(s) \quad + \quad O_2(g) \quad \longrightarrow \quad 2MgO(s)$$

3 Balance the equation.
- First, you need to add another **MgO** to the product side to balance the **O**s
- You now need to add another **Mg** on the reactant side to balance the **Mg**s
- There are two magnesium atoms and two oxygen atoms on each side – it is balanced

4 Write a balanced symbol equation.

You may be asked to include the **state symbols** when writing an equation: (aq) for aqueous solutions, (g) for gases, (l) for liquids and (s) for solids.

HT You should be able to balance equations by looking at the formulae (i.e. without drawing the atoms).

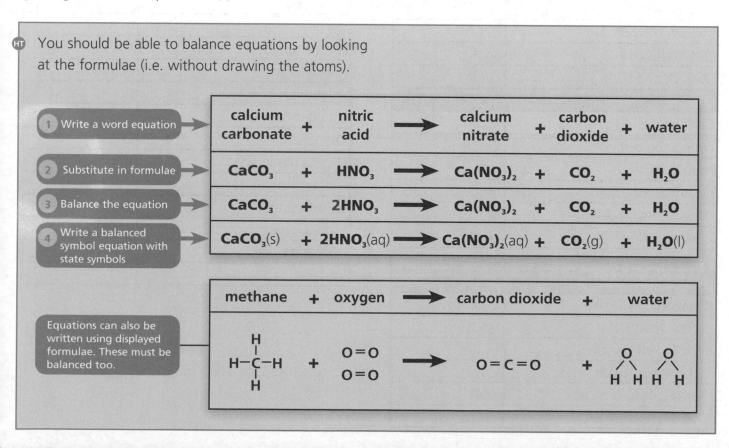

1 Write a word equation	calcium carbonate + nitric acid	→	calcium nitrate + carbon dioxide	+ water
2 Substitute in formulae	$CaCO_3$ + HNO_3	→	$Ca(NO_3)_2$ + CO_2	+ H_2O
3 Balance the equation	$CaCO_3$ + $2HNO_3$	→	$Ca(NO_3)_2$ + CO_2	+ H_2O
4 Write a balanced symbol equation with state symbols	$CaCO_3(s)$ + $2HNO_3(aq)$	→	$Ca(NO_3)_2(aq)$ + $CO_2(g)$	+ $H_2O(l)$

methane + oxygen → carbon dioxide + water

Equations can also be written using displayed formulae. These must be balanced too.

Fundamental Chemical Concepts

Common Compounds and their Formulae

Acids	
Ethanoic acid	CH_3COOH
Hydrochloric acid	HCl
Nitric acid	HNO_3
Sulfuric acid	H_2SO_4

Carbonates	
Calcium carbonate	$CaCO_3$
Copper(II) carbonate	$CuCO_3$
Iron(II) carbonate	$FeCO_3$
Magnesium carbonate	$MgCO_3$
Manganese carbonate	$MnCO_3$
Sodium carbonate	Na_2CO_3
Zinc carbonate	$ZnCO_3$

Chlorides	
Ammonium chloride	NH_4Cl
Barium chloride	$BaCl_2$
Calcium chloride	$CaCl_2$
Iron(II) chloride	$FeCl_2$
Magnesium chloride	$MgCl_2$
Potassium chloride	KCl
Silver chloride	$AgCl$
Sodium chloride	$NaCl$
Tin(II) chloride	$SnCl_2$
Zinc chloride	$ZnCl_2$

Oxides	
Calcium oxide	CaO
Copper(II) oxide	CuO
Iron(II) oxide	FeO
Magnesium oxide	MgO
Manganese(II) oxide	MnO
Sodium oxide	Na_2O
Zinc oxide	ZnO

Hydroxides	
Copper(II) hydroxide	$Cu(OH)_2$
Iron(II) hydroxide	$Fe(OH)_2$
Iron(III) hydroxide	$Fe(OH)_3$
Lithium hydroxide	$LiOH$
Potassium hydroxide	KOH
Sodium hydroxide	$NaOH$

Sulfates	
Ammonium sulfate	$(NH_4)_2SO_4$
Barium sulfate	$BaSO_4$
Calcium sulfate	$CaSO_4$
Copper(II) sulfate	$CuSO_4$
Iron(II) sulfate	$FeSO_4$
Magnesium sulfate	$MgSO_4$
Potassium sulfate	K_2SO_4
Sodium sulfate	Na_2SO_4
Tin(II) sulfate	$SnSO_4$
Zinc sulfate	$ZnSO_4$

Others	
Ammonia	NH_3
Bromine	Br_2
Calcium hydrogencarbonate	$Ca(HCO_3)_2$
Carbon dioxide	CO_2
Carbon monoxide	CO
Chlorine	Cl_2
Ethanol	C_2H_5OH
Glucose	$C_6H_{12}O_6$
Hydrogen	H_2
Iodine	I_2
Lead iodide	PbI_2
Lead(II) nitrate	$Pb(NO_3)_2$
Methane	CH_4
Nitrogen	N_2
Oxygen	O_2
Potassium iodide	KI
Potassium nitrate	KNO_3
Silver nitrate	$AgNO_3$
Sodium hydrogencarbonate	$NaHCO_3$
Sulfur dioxide	SO_2
Water	H_2O

C1: Carbon Chemistry

This module looks at:

- How crude oil is processed and used, and the problems with its exploitation.
- Fuels and complete and incomplete combustion.
- The air and pollution problems caused by burning fuels.
- Alkanes, alkenes and how alkenes can be made into polymers.
- The properties, uses and disposal of polymers.
- Chemical reactions in cooking, and the uses of food additives.
- Natural and synthetic perfumes, esters and solutions.
- How paints are made and the use of thermochromic and phosphorescent pigments.

Fossil Fuels

Crude oil, **coal** and **natural gas** are all **fossil fuels**. Fossil fuels are formed extremely slowly. It takes a very long time: millions of years! All fossil fuels are **finite**, i.e. there are limited supplies. They are described as **non-renewable** because we are using them up much faster than new supplies can be formed. This means they will eventually run out.

HT All the crude oil that is easily extracted will eventually run out and any remaining new supplies will have to be found. The search will be in more remote parts of the world and the extraction will be increasingly difficult. As oil becomes scarce, then the decision will have to be made whether to burn it as a fuel or to make petrochemicals (e.g. plastics).

Crude Oil

Crude oil is found trapped in the Earth's crust. To release the oil, a hole is drilled through the rock. If the oil is under pressure, it will flow out. If it is not under pressure, it has to be pumped out.

When crude oil is extracted it is a **thick**, **black**, **sticky liquid**. It is transported to a **refinery** through a pipeline or in oil tankers. This is a dangerous procedure: if the oil accidentally spills into the sea, it can have a devastating effect on wildlife. Oil spills or **slicks** float on the sea's surface. The toxic oil can coat the feathers of sea birds preventing them from floating or flying and may kill them. If an oil slick washes ashore it can damage the beaches leading to large clean-up operations. Detergents are often used to disperse oil slicks or remove oil from beaches but detergents are also toxic to wildlife.

Fractional Distillation

Crude oil is a mixture of many **hydrocarbons**. A hydrocarbon is a molecule that contains only **carbon** and **hydrogen** atoms. Different hydrocarbons have different boiling points. This means crude oil can be separated into useful products or **fractions** (parts) by **heating** it in a process called **fractional distillation**.

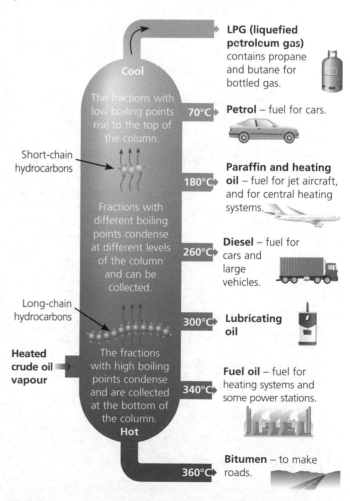

Cool

The fractions with low boiling points rise to the top of the column.

Short-chain hydrocarbons

Fractions with different boiling points condense at different levels of the column and can be collected.

Long-chain hydrocarbons

Heated crude oil vapour

The fractions with high boiling points condense and are collected at the bottom of the column.

Hot

70°C

180°C

260°C

300°C

340°C

360°C

LPG (liquefied petroleum gas) contains propane and butane for bottled gas.

Petrol – fuel for cars.

Paraffin and heating oil – fuel for jet aircraft, and for central heating systems.

Diesel – fuel for cars and large vehicles.

Lubricating oil

Fuel oil – fuel for heating systems and some power stations.

Bitumen – to make roads.

Forces Between Molecules

The **atoms** in a hydrocarbon molecule are strongly held together by the bonds between them, for example:

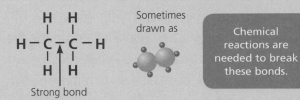

Strong bond

Sometimes drawn as

Chemical reactions are needed to break these bonds.

All hydrocarbon molecules have forces of attraction between them called **intermolecular forces**, but they are only weak. However, the longer the hydrocarbon molecule is, the stronger the intermolecular forces are, for example:

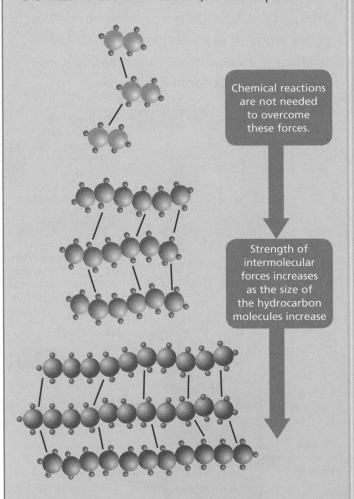

Chemical reactions are not needed to overcome these forces.

Strength of intermolecular forces increases as the size of the hydrocarbon molecules increase

The bonds between the carbon and hydrogen atoms within a hydrocarbon are stronger than the forces between hydrocarbon molecules.

Separating Hydrocarbons

When a hydrocarbon is heated, its molecules move faster and faster until the intermolecular forces are broken.

Small molecules have very small forces of attraction between them and are easy to break by heating. This means that hydrocarbons with small molecules are volatile liquids or gases with low boiling points, for example:
- Methane, CH_4, has a boiling point of -164°C.
- Ethane, C_2H_6, has a boiling point of -89°C.

Short-chain Hydrocarbons

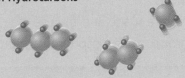

Large molecules have many more of these small forces between them, resulting in an overall large force of attraction. This force is more difficult to break by heating and hydrocarbons with large molecules are thick, viscous liquids or waxy solids with higher boiling points, for example:
- Octane, C_8H_{18} has a boiling point of 126°C.
- Decane, $C_{10}H_{22}$ has a boiling point of 174°C.

Long-chain Hydrocarbons

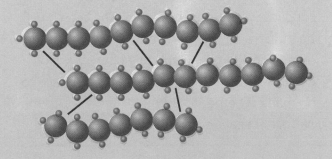

It is the differences in their boiling points which enables a mix of hydrocarbons (e.g. crude oil) to be separated by the process of fractional distillation.

The Fractions

Each fraction consists of a mixture of hydrocarbons whose boiling points fall within a particular **range**.

This table shows the main fractions obtained through the industrial fractional distillation of crude oil, and their approximate boiling ranges.

Fraction	Boiling Range
Refinery gases	up to 25°C
Petrol	40–100°C
Paraffin and heating oil	150–250°C
Diesel	220–350°C
Lubricating oil	over 350°C
Fuel oil	over 400°C
Bitumen	over 400°C

Cracking

Hydrocarbon molecules can be described as **alkanes** or **alkenes**, depending on whether or not they have a carbon–carbon double bond present (see page 52).

Cracking converts **large** alkane molecules into **smaller**, more useful, alkane and alkene molecules. The alkene molecules obtained can be used to make polymers, which have many uses (see pages 54–55). The smaller alkane molecules obtained are usually blended to make petrol, which is in huge demand. You may be asked to answer questions about the supply and demand of crude oil fractions in the exam – the information will be given to you to interpret.

To take place, cracking needs a **catalyst** and a **high temperature**. In the laboratory, cracking is carried out using the apparatus shown below.

Long-chain hydrocarbon — heat / catalyst → Short-chain hydrocarbons

There is not enough petrol in crude oil to meet demand. Therefore, cracking is used to convert parts of crude oil that cannot be used into additional petrol.

Crude oil and natural gas are found in many parts of the world. The UK is now dependent on oil and gas from some politically unstable countries, which could cause supply problems in the future.

A Cracking Reaction

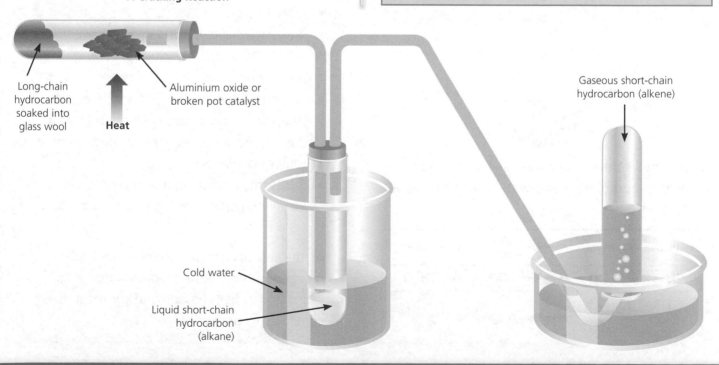

47

Choosing a Fuel

Some or all of the following factors should be taken into account when choosing a fuel for a specific purpose:

- **Energy value** – how much energy do you get from a measured amount of fuel?
- **Availability** – is the fuel easy to obtain?
- **Storage** – how easy is it to store the fuel? (e.g. petrol is more difficult to store than coal.)
- **Cost** – how much fuel do you get for your money?
- **Toxicity** – is the fuel (or its combustion products) poisonous?
- **Pollution** – do the combustion products pollute the atmosphere? (e.g. acid rain or the greenhouse effect)
- **Ease of use** – is it easy to control and is special equipment needed?

You may be asked to look at data about a number of fuels and decide which one is the best for a particular purpose.

Burning Fuels (Combustion)

When fuels burn, useful energy is released as heat. Chemists call this **combustion**. Fuels are substances that react with oxygen in the air. Complete combustion needs a plentiful supply of oxygen.

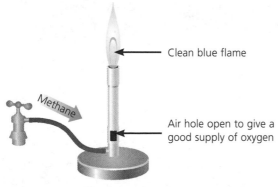

Clean blue flame

Methane

Air hole open to give a good supply of oxygen

When a hydrocarbon, like methane, is burned in air, only carbon dioxide and water (hydrogen oxide) are formed.

| methane | + | oxygen | → | carbon dioxide | + | water |

$$CH_4 + 2O_2 \longrightarrow CO_2 + 2H_2O$$

Evaluating a Fuel

Choosing a fuel to use for a particular job requires a careful study of available information.

In your exam you may be asked to evaluate the use of different fossil fuels using given data, e.g. tables, graphs, pie charts. The following are examples of the sort of fuels that might be considered.

Methane (CH_4)
- Colourless gas.
- Burns to form carbon dioxide and water.
- Non-toxic (but it is a greenhouse gas).
- Readily available through normal gas supplies.
- Not easy to store.
- 1g of methane produces 55.6kJ of energy when completely burned.

Butane (C_4H_{10})
- Easier to store and transport than methane.
- Burns in the same way as methane.
- Used as camping gas.
- Only 26.9kJ of energy is produced from 1 gram when it is burned.

Coal
- Easy to store.
- Readily available, not very expensive and releases quite a lot of energy when burned.
- Main problem is pollution and in particular the sulfur dioxide gas (that leads to acid rain) it produces, along with smoke and other pollutants.
- Most major populated areas of the UK allow only smokeless coal to be burned as a fuel.

As the world's population increases and more countries become industrialised (e.g. China and India), the demand for fossil fuels continues to grow.

Detecting the Products of Combustion

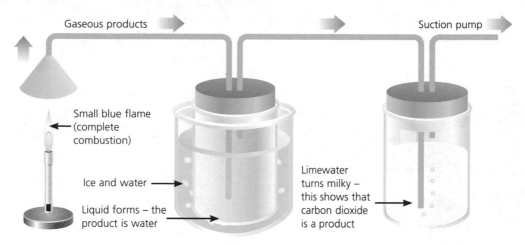

Gaseous products → → Suction pump →

Small blue flame (complete combustion)

Ice and water →

Liquid forms – the product is water

Limewater turns milky – this shows that carbon dioxide is a product

Incomplete Combustion

If a fuel burns without sufficient oxygen, e.g. in a room with poor ventilation or when a gas appliance needs servicing, then **incomplete combustion** takes place and **carbon monoxide** (a poisonous gas) can be formed. For example, the incomplete combustion of methane:

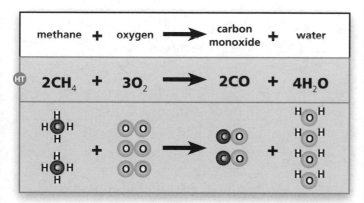

methane + oxygen → carbon monoxide + water

HT $2CH_4 + 3O_2 \longrightarrow 2CO + 4H_2O$

If there is **very little oxygen** available, **carbon** (soot) is produced instead. For example, the burning of methane when very little oxygen is available:

methane + oxygen → carbon + water

HT $CH_4 + O_2 \longrightarrow C + 2H_2O$

Although incomplete combustion releases some energy, much more is released when complete

combustion takes place. The following are also advantages of making sure a fuel burns completely:

• Less soot is produced.
• No poisonous carbon monoxide gas is produced.

A blue flame on a Bunsen burner transfers more energy than a yellow flame because it involves complete combustion. The yellow flame shows that incomplete combustion is taking place.

You should be able to write the word equations for complete and incomplete combustion if you are given the name of the fuel. You may not be given all the names of some of the reactants and products so remember the following points:

• Combustion is a reaction with oxygen.
• Complete combustion of a hydrocarbon produces water and carbon dioxide.
• Incomplete combustion of a hydrocarbon produces water and carbon monoxide.
• Incomplete combustion when there is only a small amount of oxygen produces water and carbon.

HT You should be able to write a balanced symbol equation for complete and incomplete combustion given the formula of the fuel. Remember:
• The formula of oxygen is O_2.

The Changing Atmosphere

The Earth's atmosphere has not always been the same as it is today. It has gradually changed over billions of years.

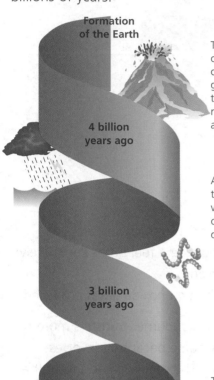

Formation of the Earth

The earliest atmosphere contained ammonia and carbon dioxide. These gases came from inside the Earth and were often released through the action of volcanoes.

4 billion years ago

As the temperature of the planet fell, the water vapour in the atmosphere condensed to form the oceans and seas.

3 billion years ago

The evolution of plants meant that photosynthesis started to reduce the amount of carbon dioxide and increase the amount of oxygen in the atmosphere.

2 billion years ago

1 billion years ago

Now

Clean air contains about:
- **78% nitrogen**
- **21% oxygen**
- **1% other gases, including 0.035% carbon dioxide**.

It also contains varying amounts of **water vapour** and some **polluting gases**.

Respiration and Photosynthesis

All living things respire. They take in oxygen and give out carbon dioxide. This decreases the oxygen levels and increases the carbon dioxide levels in the air.

Animals respire all the time; plants also respire all the time, but during the day plants also photosynthesise. This is the opposite of respiration: they take in carbon dioxide and release oxygen. Photosynthesis and respiration balance out, so the levels of carbon dioxide and oxygen in the present-day air remain fairly constant. The levels of nitrogen in the air also stay fairly constant.

HT Theories of Atmospheric Change

Explanations about how our planet and its atmosphere have evolved are scientists' best efforts at interpreting all the available evidence. Below is one suggested explanation. But remember, it is only a theory!

A hot volcanic Earth would have released various gases into the atmosphere, just as volcanoes do nowadays. These gases would probably have included water vapour and carbon dioxide (and small amounts of ammonia, methane and sulfur dioxide).

As the Earth cooled down, its surface temperature would have gradually fallen below 100°C and water vapour would have condensed to form the oceans.

The levels of carbon dioxide started to decrease as it dissolved in the oceans. Further reduction in carbon dioxide levels came about when algae and simple plants evolved and used it for photosynthesis. This process also led to the increase in oxygen levels in the atmosphere.

Ammonia in the early atmosphere was converted into nitrogen by the action of bacteria. Nitrogen is very unreactive and so the level of nitrogen in the air has gradually increased.

Air Pollution

Pollutant gases are formed from:

• the burning of fossil fuels

• incomplete combustion in a car engine.

In
• Air – mostly nitrogen and oxygen.
• Hydrocarbon fuel such as petrol or diesel.

Out
• The normal products of combustion.
• **Carbon monoxide** – poisonous gas made when the fuel does not burn completely.
• **Oxides of nitrogen** – formed inside the internal combustion engine.

The table below includes information on how pollutant gases are produced, and what environmental problems they lead to. In your exam you may be asked to use information such as this to make judgements about the effects of air pollution, e.g. how it affects people's health.

Source	Gas Produced
Car exhausts	• Carbon monoxide (poisonous). • Nitrogen dioxide (leads to acid rain). • Unburned hydrocarbons (creates smog).
Oxides of nitrogen	• Acid rain and photochemical smog.
Aerosols	• CFCs (damage the ozone layer).

Burning fossil fuels (which are all carbon compounds) increases the amount of carbon dioxide in the atmosphere. However, this carbon dioxide can be used for photosynthesis.

It is important to reduce air pollution as much as possible because it can damage our surroundings and can adversely affect people's health. One way to remove carbon monoxide from car exhausts is to fit a **catalytic converter**. The catalyst causes the carbon monoxide to react, producing carbon dioxide.

Acid Rain

When coal or oils are burned, the sulfur impurities produce sulfur dioxide. Sulfur dioxide (and nitrogen dioxide) dissolves in water to produce acid rain. Acid rain can:

• erode stonework and corrode metals

• make rivers and lakes acidic and kill aquatic life

• kill plants.

Human Influence on the Atmosphere

Until relatively recently, the balance between adding and removing carbon dioxide from the atmosphere had remained constant. The levels of carbon dioxide and oxygen were maintained by photosynthesis and respiration. However, three important factors have upset the balance:

1 Excessive burning of fossil fuels is increasing the amount of carbon dioxide in the atmosphere.

2 Deforestation on large areas of the Earth's surface means the amount of photosynthesis is reduced so less carbon dioxide is removed from the atmosphere.

3 The increase in world population has directly and indirectly contributed to factors **1** and **2**.

Oxygen and nitrogen from the air react together in the high temperature of a car engine to form oxides of nitrogen.

To help reduce the amount of pollutants being put into the atmosphere, catalytic converters are fitted to cars to convert the carbon monoxide in exhaust gases to the less harmful carbon dioxide.

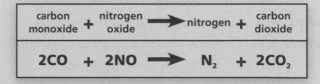

carbon monoxide	+	nitrogen oxide	→	nitrogen	+	carbon dioxide
2CO	**+**	**2NO**	**→**	**N$_2$**	**+**	**2CO$_2$**

Hydrocarbons

Hydrocarbons are molecules that contain hydrogen and carbon atoms only.

You need to remember that:
- carbon atoms can make four bonds each
- hydrogen atoms can make one bond each.

Alkanes

When a hydrocarbon contains **single covalent bonds** only, it is called an **alkane**. The name of an alkane always ends in -**ane**.

This table shows the displayed and molecular formulae for the first four members of the alkane series.

Alkane	Displayed Formula	Molecular Formula
Methane		CH_4
Ethane		C_2H_6
Propane		C_3H_8
Butane		C_4H_{10}

Alkenes

The **alkenes** are another form of hydrocarbon. They are very similar to the alkanes except that they contain **one carbon–carbon double covalent bond** between two adjacent carbon atoms. A double bond contains two shared pairs of electrons. The name of an alkene always ends in -**ene**.

This table shows the displayed and molecular formulae for the first three members of the alkene series.

Alkene	Displayed Formula	Molecular Formula
Ethene		C_2H_4
Propene		C_3H_6
Butene		C_4H_8

A simple test to distinguish between alkenes and alkanes is to add bromine water. Alkenes react with bromine water (orange) and decolourise it (colourless). Alkanes have no effect on bromine water.

HT Alkanes contain only single covalent bonds between the carbon atoms – they are described as **saturated** hydrocarbons. (They have the maximum number of hydrogen atoms per carbon atom in the molecule.)

Alkenes contain at least one carbon–carbon double covalent bond. This means that the carbon atom is not bonded to the maximum number of hydrogen atoms. Alkenes are therefore described as being **unsaturated**.

Bromine water (orange) turns colourless when shaken with an alkene. This is an **addition reaction** as the bromine adds on to the alkene molecule to make a **colourless dibromo compound**, e.g.

$$C_2H_4 + Br_2 \longrightarrow C_2H_4Br_2 \text{ (1,2-dibromoethane)}$$

Polymerisation

The alkenes made by cracking can be used as **monomers**. Monomers are small molecules that can be reacted together to produce **polymers**. These are very large molecules, some of which make up plastics.

Alkenes are very good at joining together, and when they do so without producing another substance we call it **polymerisation**.

This process, e.g. the formation of poly(ethene) from ethene, requires **high pressure** and a **catalyst**.

The name of a polymer is made from the name of its monomer, e.g. ethene makes poly(ethene) and you can work out what the monomer is from the name of the polymer, e.g. poly(propene) is made from propene.

The small alkene molecules are called monomers.

Their carbon–carbon double bonds are easily broken.

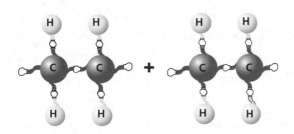

Large numbers of molecules can therefore be joined in this way. The resulting long-chain molecule is a polymer: poly(ethene) or polythene.

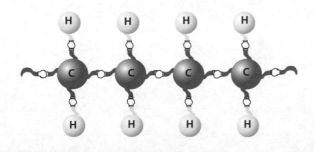

Consider the displayed formula of ethene and poly(ethene):

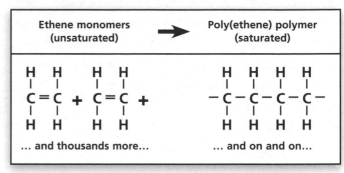

A more compact way of writing this reaction uses the standard way of displaying a polymer formula:

$$n \left[\begin{array}{c} H \quad H \\ C=C \\ H \quad H \end{array} \right] \rightarrow \left[\begin{array}{c} H \; H \\ -C-C- \\ H \; H \end{array} \right]_n$$

HT Addition polymerisation involves the reaction of many unsaturated monomer molecules, i.e. alkenes, to form a saturated polymer. You will be expected to be able to construct the displayed formula of:

- a **polymer** given the displayed formula of a monomer, e.g. propene monomer to poly(propene) polymer:

$$\begin{array}{c} H \quad CH_3 \\ C=C \\ H \quad H \end{array} \rightarrow \left[\begin{array}{c} H \; CH_3 \\ -C-C- \\ H \; H \end{array} \right]$$

- a **monomer**, given the displayed formula of a polymer, e.g. poly(propene) polymer to propene monomer:

$$\left[\begin{array}{c} H \; CH_3 \\ -C-C- \\ H \; H \end{array} \right] \rightarrow \begin{array}{c} H \quad CH_3 \\ C=C \\ H \quad H \end{array}$$

Polymers

Polymers (plastics) have many properties that make them useful. You should be able to use these terms to explain why a certain plastic is used for a particular job. Properties of plastics are listed below:

- Can be easily moulded into shape
- Waterproof
- Electrical insulator
- Non-biodegradable
- Lightweight
- Flexible
- Can be printed on
- Unreactive
- Can be coloured
- Heat insulator
- Transparent
- Tough

Uses for Polymers

Different plastics have different properties, which results in them having different uses:

Polymer	Properties	Uses
Polythene or poly(ethene)	• Lightweight • Flexible • Easily moulded	• Plastic bags • Moulded containers
Polystyrene (expanded polystyrene)	• Lightweight • Poor conductor of heat	• Insulation • Damage protection in packaging
Nylon	• Lightweight • Waterproof • Tough	• Clothing • Climbing ropes
Polyester	• Lightweight • Waterproof • Tough	• Clothing • Bottles

Outdoor Clothing

Outdoor clothing, such as a jacket, needs to be waterproof to keep the wearer dry. Nylon is an excellent material to use to make outdoor clothing because it is:

- lightweight
- tough
- waterproof (but it does not let water vapour escape, so it could be uncomfortable to wear if the wearer becomes hot and starts to perspire)
- able to block ultraviolet (UV) light (harmful sunlight).

Gore-Tex®

Gore-Tex® is a breathable material made from nylon. It has all of the advantages of nylon, but it is also treated with a material that allows perspiration (water vapour) to escape whilst preventing rain from getting in. This is far more comfortable for people who lead an active outdoor life, as it prevents them from getting wet when they perspire.

HT Gore-Tex® has a membrane of polyurethane or poly(tetrafluoroethane) (PTFE), sandwiched between two layers of nylon fibres. The laminated PTFE has very tiny holes that allows water vapour to pass through but that are too small to allow liquid water to pass through. The PTFE laminate is too weak and fragile to be used on its own.

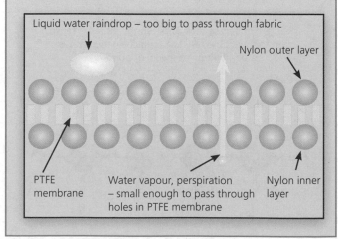

Liquid water raindrop – too big to pass through fabric

Nylon outer layer

PTFE membrane

Water vapour, perspiration – small enough to pass through holes in PTFE membrane

Nylon inner layer

Structure of Plastics

Polymers (plastics), such as poly(chloroethene) (PVC), consist of a tangled mass of very long-chain molecules, in which the atoms are held together by strong covalent bonds. The properties of a plastic depend on its structure.

Plastics that have weak forces between polymer molecules have low melting points and can be stretched easily as the polymer molecules can slide over one another.

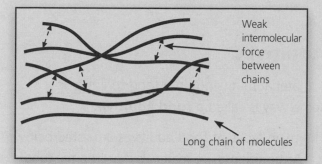

Weak intermolecular force between chains

Long chain of molecules

Plastics that have strong forces between the polymer molecules (covalent bonds or cross-linking bridges) have high melting points, are rigid and cannot be stretched.

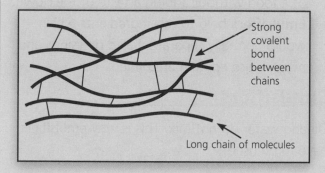

Strong covalent bond between chains

Long chain of molecules

Disposal of Plastics

As we have seen, plastics have many different uses. As it is such a convenient material, we produce a large amount of plastic waste. This can be difficult to dispose of and can sometimes be seen as litter in the streets. There are various ways of disposing of plastics. Unfortunately, some of them have a negative impact on the environment:

- Using **landfill sites** is a problem because most plastics are non-biodegradable. This means microorganisms have no effect on them and they will not decompose and rot away. Throwing plastics into landfill sites results in the waste of a valuable resource and, because of the volume of waste produced, landfill sites get filled up very quickly, which is also a waste of land.
- **Burning** plastics produces air pollution and also wastes valuable resources. The production of carbon dioxide contributes to the greenhouse effect which results in global warming. Some plastics cannot be burned at all because they produce toxic fumes. For example, burning poly(chloroethene) or PVC as it is more commonly known, produces hydrogen chloride gas.
- **Recycling** plastics is an option which prevents resources being wasted. However, different types of plastic need to be recycled separately. Sorting them into groups to be recycled can be difficult and very time-consuming.

Research is being carried out on the development of biodegradable plastics to help reduce the impact that the disposal of plastics has on the environment.

Soluble plastics make disposal easy. For example, the plastic case of a dishwasher tablet is disposed of when it dissolves in contact with hot water, releasing detergent into the machine.

Cooking Food

Cooking food causes a **chemical change** to take place. When a chemical change occurs:

- new substances are formed from old ones
- there may be a change in mass when a gas is released
- there is often a substantial energy change, e.g. a rise or fall in temperature
- the change cannot be reversed easily.

Cooking Eggs and Meat

Eggs and meat contain lots of protein. The protein molecules change shape when they are heated. This is called **denaturing**.

> **HT** The texture of eggs and meat changes when they are cooked because the protein molecules change shape permanently.
>
> #### Potatoes and Vegetables
> Potatoes and other vegetables are plants; their cells have a rigid cell wall. During cooking, the heat breaks down this cell wall, starch is released and it becomes much softer. The starch grains swell up and spread out, and the potato is now much easier to digest.

Baking Powder

Baking powder contains sodium hydrogencarbonate. When this is heated, it breaks down (decomposes) to make sodium carbonate and water, and carbon dioxide gas is given off.

The word equation for the decomposition of sodium hydrogencarbonate is:

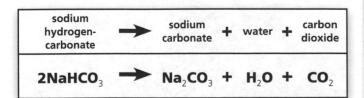

sodium hydrogen-carbonate	→	sodium carbonate	+ water	+ carbon dioxide
$2NaHCO_3$	→	Na_2CO_3	+ H_2O	+ CO_2

Baking powder is added to cake mixture because the carbon dioxide gas given off when it is heated causes the cake to **rise**.

You can test for the presence of carbon dioxide using **limewater** (**calcium hydroxide solution**). If carbon dioxide is present, the limewater turns **milky**.

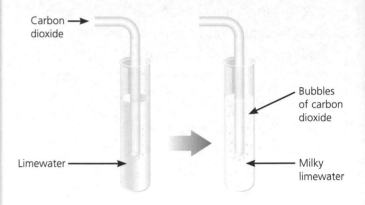

Additives

A material that is put in a food to improve it in some way is called a **food additive**.

The main types of food additives are listed below:

- **Antioxidants** are materials that stop the food reacting with oxygen in the air. They are usually added to foods that contain fats or oils, e.g. bacon.
- **Food colours** improve the appearance of food.
- **Flavour enhancers** help bring out the flavour of a food without adding a taste of their own.
- **Emulsifiers** help to mix ingredients which would normally separate. Salad dressings and mayonnaise contain emulsifiers.

Emulsifiers

Oil and water do not mix. This is why emulsifiers have to be used.

The molecules in an emulsifier have two ends: one end likes to be in water (**hydrophilic**) and the other end likes to be in oil (**hydrophobic**). The emulsifier joins the droplets together and keeps them mixed.

> **HT** The **hydrophilic** end of the emulsifier molecule bonds to the **polar water molecules**. The **hydrophobic** end of the emulsifier molecule bonds to the **non-polar oil molecules**.

Perfumes

Smells are made of molecules which travel up your nose and stimulate sense cells.

A perfume must smell nice. In addition, it must:
- evaporate easily – so it can travel to your nose
- not be toxic – so it does not poison you
- not irritate – otherwise it would be uncomfortable on the skin
- not dissolve in water – otherwise it would wash off easily
- not react with water – otherwise it would react with perspiration.

There are many kinds of perfume. Some come from **natural** sources, such as plants and animals. Perfumes can also be **manufactured**. If they are manufactured they are known as **synthetic** perfumes.

Esters are a common family of compounds used as synthetic perfumes. An ester is made by reacting an alcohol with an organic acid. This produces an ester and water. A simple ester, ethyl ethanoate, is made by adding ethanoic acid to ethanol (see below).

Perfumes and cosmetics need to be tested to make sure they are not harmful. This testing is sometimes done on animals. Some people are not happy about this. They argue that it is cruel to animals, and pointless because animals do not have the same body chemistry as humans and so results of the tests may not be useful. However, the tests could be useful to prevent humans from being harmed. The testing of cosmetics on animals is now banned in the EU.

HT Perfumes are **volatile**: they evaporate easily.

The molecules in a drop of perfume are held together by weak intermolecular forces of attraction. The molecules that escape have lots of energy and easily overcome the weak attraction to the other molecules in the liquid.

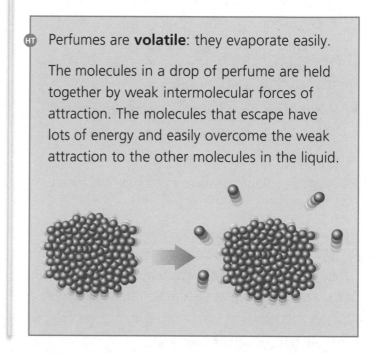

Making a Simple Ester

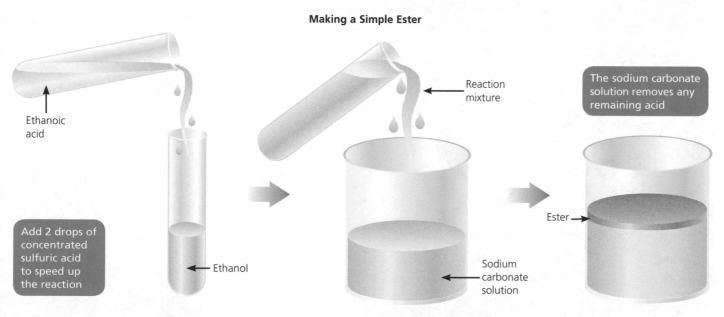

Ethanoic acid

Add 2 drops of concentrated sulfuric acid to speed up the reaction

Ethanol

Reaction mixture

Sodium carbonate solution

The sodium carbonate solution removes any remaining acid

Ester

Solvents

Below are some words that are used to describe substances (together with their definitions):

- **Soluble substances** are substances that dissolve in a liquid, e.g. nail varnish is soluble in ethyl ethanoate.
- **Insoluble substances** are substances that do not dissolve in a liquid, e.g. nail varnish is insoluble in water.
- A **solvent** is the liquid into which a substance is dissolved, e.g. ethyl ethanoate is a solvent. (An ester can be used as a solvent.)
- The **solute** is the substance that gets dissolved, e.g. the nail varnish is a solute.
- A **solution** is what you get when you mix a solvent and a solute. The mixture does not separate out.

Nail varnish colours dissolve in nail varnish remover (a solvent).

You may be asked to put solvents in order of most effective to least effective in removing nail varnish, or to suggest which solvent you could use to dissolve magnesium chloride (a salt).

You may be asked to decide how good a solvent is from information collected in an experiment, such as in this table:

Solvent	Nail Varnish	Sodium Chloride (Salt)
Water	Does not dissolve	Very soluble
Ethanol	Dissolves in 15 seconds	Slightly soluble
Ethyl ethanoate	Dissolves in 3 seconds	Insoluble
Propanone	Dissolves in 2 seconds	Insoluble

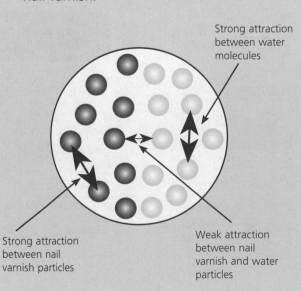

(HT) Water will not dissolve nail varnish because:

- the attraction between water molecules is stronger than the attraction between water molecules and the particles in nail varnish
- the attraction between the particles in nail varnish is stronger than the attraction between water molecules and particles in nail varnish.

Strong attraction between water molecules

Strong attraction between nail varnish particles

Weak attraction between nail varnish and water particles

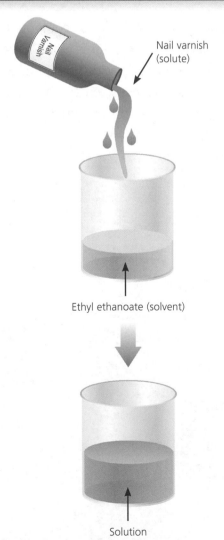

Nail varnish (solute)

Ethyl ethanoate (solvent)

Solution

Paint

Paint can be used to protect or to decorate a surface. It is a special mixture of different materials, and it is called a **colloid**. In a colloid, fine solid particles are well mixed (dispersed) with liquid particles but they are not dissolved.

Paint is a mixture of:
* a **pigment** – a substance that gives the paint its colour
* a **binding medium** – an oil that sticks the pigment to the surface it is being painted onto
* a **solvent** – dissolves the thick binding medium and makes it thinner and easier to coat the surface.

The paint coats a surface with a thin layer and the solvent evaporates away as the paint dries. The solvent in **emulsion** paint is **water**. In **oil-based** paints, the pigment is dispersed in an **oil** (the binding medium). Often, there is a solvent present that dissolves the oil.

Special Pigments

Thermochromic pigments change colour when they are heated or cooled. These pigments can be used in:
* cups and kettles to warn that they are hot
* mood rings that change colour as your body temperature changes
* baby feeding spoons to show that the food is not too hot
* bath toys to show that the water is the correct temperature for a baby.

Phosphorescent pigments glow in the dark. They absorb and store energy and release it slowly as light when it is dark. The paint on some watch dials contains phosphorescent pigments.

More about Paint

The particle size of the solids in a colloid must be very small so they stay scattered throughout the mixture. If the particles were too big, they would start to settle down to the bottom.

An oil-based paint such as a gloss paint dries in two stages:
1. The solvent **evaporates** away.
2. The oil-binding medium reacts with oxygen in the air as it dries to form a hard layer. This is an **oxidation** reaction.

More about Pigments

There are only a few thermochromic pigments. To increase the range of colours, they can be mixed with ordinary pigments in acrylic paints.

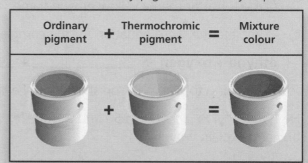

Ordinary pigment	+	Thermochromic pigment	=	Mixture colour

Thermochromic pigments change to colourless as they get hotter and so the paint changes from the mixture colour to the ordinary pigment colour.

Cold → Hot

The first 'glow in the dark' paints were made using radioactive materials as pigments. They were used to paint the dials on aircraft instrument panels and the first luminous watches. However, the people who painted with these pigments were exposed to too much radiation and some of them developed cancer as a result. Phosphorescent pigments are not radioactive, so they are much safer to use.

1 **a)** Explain why diesel fuel is non-renewable. [2]

b) An oil company wants to build an oil tanker terminal near a large sea bird colony.

Write about the advantages and disadvantages of building a terminal in this area. [6]

✍ *The quality of written communication will be assessed in your answer to this question.*

c) Look at the formulae of these molecules:

A CH_4 **B** C_3H_6 **C** C_4H_{10} **D** C_2H_5OH **E** C_3H_8

     i) Which molecule is not a hydrocarbon? [1]
     ii) Which molecule contains 9 atoms? [1]

d) Benzene is found in crude oil. It has a boiling point of 81°C. Suggest which of the fractions of crude oil will contain benzene after distillation. Use the information in the table to help you. [1]

Fraction	Refinery gases	Petrol	Paraffin/ heating oil	Diesel	Lubricating oil	Fuel oil	Bitumen
Boiling range (°C)	Up to 25	40–100	150–250	220–350	Over 350	Over 400	Over 400

2 **a)** Suggest two reasons why methane gas is used in many homes as a fuel for cooking. [2]

b) Ethyne is a hydrocarbon gas that is used as a fuel in welding torches. It burns with a cold yellow flame but when oxygen is added it burns with a very hot, blue flame. Explain why a blue ethyne flame is hotter than a yellow ethyne flame. [2]

c) Complete the word equation for the complete combustion of ethyne.

ethyne + oxygen ➔ _____ + _____ [1]

3 Joe is planning to walk the Yorkshire Three Peaks. He wants to buy a suitable jacket.

He finds a nylon jacket for £50 and a Gore-Tex® jacket for £65. Suggest which jacket Joe should buy and explain why. [2]

4 **a)** An ester is made by this reaction: **ethanol + ethanoic acid ➔ ethyl ethanoate + water**
Give the name of a reactant in this reaction. [1]

b) A scientist has developed a new paint mixture and she now wants to try different colours. What should she change to get a range of different colours? [1]

5 David and Lisa measure the concentration of sulfur dioxide gas in the air starting at the centre of town and moving out into the countryside. Their results are shown in the table.

Distance from town centre (km)	0	2	4	6	8	10
Sulfur dioxide level (µg/m³)	122	113	95	83	116	82

David claims the results show that the greater distance you are from the town centre the less air pollution there is. Comment on David's claim and discuss how they could improve their evidence. [3]

HT **6** Write the formula equation for the complete combustion of methane. [2]

7 Explain how cutting down large areas of rainforest in South America affects the Earth's atmosphere. [5]

8 An engineer is having problems with her high temperature fuel burner. Nitrogen oxides are released in the exhaust, polluting the air even though there is no nitrogen in the hydrocarbon fuel. Explain why nitrogen oxides are made and suggest how the problem can be solved. [3]

C2: Chemical Resources

This module looks at:

- The structure of the Earth, volcanoes and the theory of plate tectonics.
- How rocks are used as the raw materials for the construction industry.
- Metal properties and alloys, and how metals are extracted from ores.
- The materials used in car making, and comparisons between aluminium and iron.
- The manufacture of ammonia and the conditions used in the Haber process.
- Acids, bases and neutralisation reactions.
- The benefits and disadvantages of using chemical fertilisers.
- The electrolysis of sodium chloride as a source of important materials.

Structure of the Earth

The Earth is nearly spherical and has a layered structure as shown below.

The thickness of the thin, rocky **crust** varies between 10km and 100km. **Oceanic** crust lies beneath the oceans. **Continental** crust forms the continents.

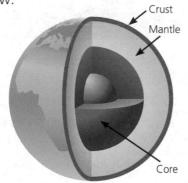

Crust

Mantle

Core

The **mantle** extends almost halfway to the centre of the Earth. It has a higher density than rock in the crust, and has a different composition.

The **core** accounts for over half of the Earth's radius. It is mostly made of iron.

It is difficult to collect information about the structure of the Earth. The deepest mines and deepest holes drilled into the crust have penetrated only a few kilometres. Scientists have to rely on studying the seismic waves (vibrations) caused by earthquakes and man-made explosions.

Movement of the Lithosphere

The Earth's **lithosphere** is the relatively cold, rigid, outer part of the Earth, consisting of the crust and outer part of the mantle. The top of the lithosphere is 'cracked' into several large interlocking pieces called **tectonic plates.**

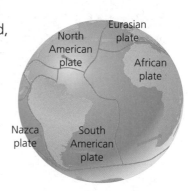

North American plate

Eurasian plate

African plate

Nazca plate

South American plate

Many ideas have been used to explain how the Earth's surface behaves, but most scientists now accept the theory of plate tectonics. It fits with a wide range of evidence and many scientists have discussed and tested it.

The plates sit on top of the mantle because they are less dense than the mantle itself. Although there does not appear to be much going on, the Earth and its crust are very dynamic. They move very slowly, at speeds of about 2.5 cm a year. Plates can move apart from, towards, or slide past, each other. This movement causes **earthquakes** and **volcanoes** at the boundaries between plates. It has taken millions of years for the continents to have moved to where they are today.

Volcanoes

Volcanoes form where molten rock can find its way through to the Earth's surface, usually at plate boundaries or where the crust is weak.

Volcanoes can give out lava that can move fast and is very runny, or erupt thick lava violently with disastrous effects. Living near a volcano can be very dangerous, but people often choose to live there because volcanic soil is very fertile.

Geologists study volcanoes to help understand the structure of the Earth and also to help predict when eruptions will occur to give an early warning for people who live nearby. However, they still cannot predict with 100% certainty.

Volcanoes (cont)

Igneous rock is formed when molten rock cools down. Igneous rocks are hard and have interlocking crystals:

- Large crystals form when molten rock cools slowly.
- Small crystals form when molten rock cools quickly.

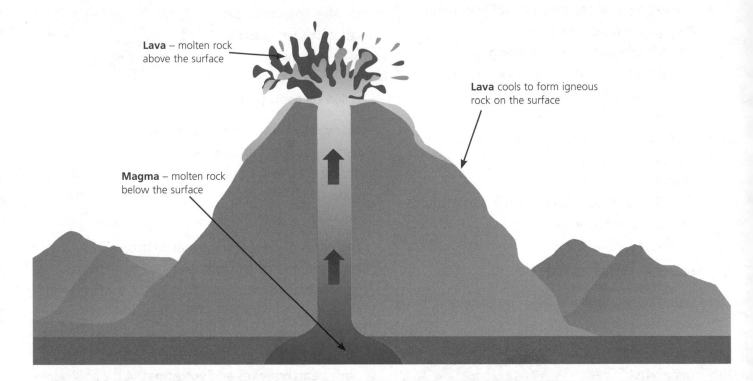

Lava – molten rock above the surface

Lava cools to form igneous rock on the surface

Magma – molten rock below the surface

Magma

The different compositions of magma affect the type of rock that will form, and the type of eruption.

Iron-rich basalt magma is quite runny and fairly safe in comparison with silica-rich rhyolite, which is thicker, and treacle-like. The volcanoes that have thicker magma can erupt violently.

Geologists are getting better at forecasting volcanic eruptions but they still cannot predict with 100% certainty.

What Causes Plates to Move?

In the zone below the lithosphere and above the core, the mantle is relatively cold and rigid. At greater depths, the mantle is hot, non-rigid and able to flow. Within the mantle, at these greater depths, are **convection currents** which are driven by heat released from radioactivity. These convection currents in the semi-rigid mantle transfer energy to the plates which then move slowly.

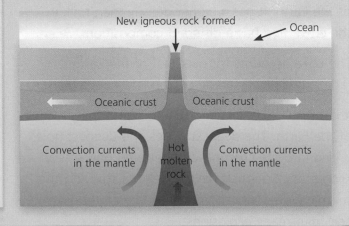

New igneous rock formed

Ocean

Oceanic crust

Oceanic crust

Convection currents in the mantle

Hot molten rock

Convection currents in the mantle

Effects of Plate Collision

While some plates around the world are moving apart, others are moving **towards** each other. An oceanic plate 'dips down' when it collides with a continental plate and slides under it, because oceanic crust has a higher density than continental crust. This is called **subduction**.

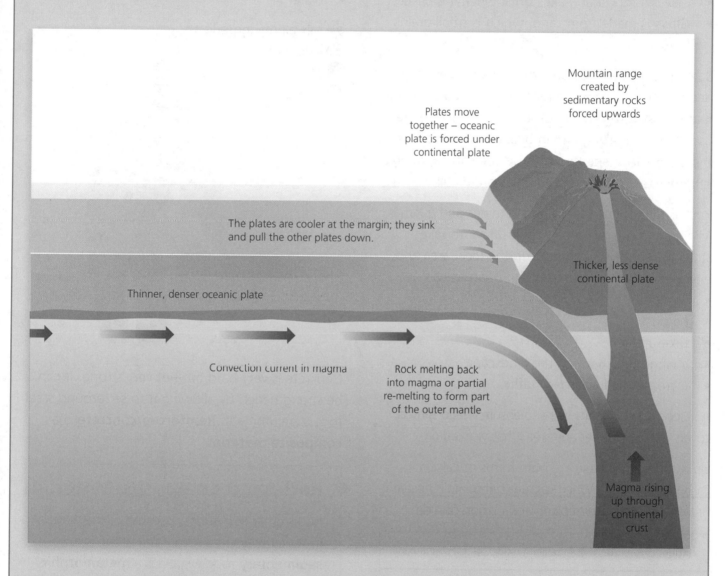

Plates move together – oceanic plate is forced under continental plate

Mountain range created by sedimentary rocks forced upwards

The plates are cooler at the margin; they sink and pull the other plates down.

Thicker, less dense continental plate

Thinner, denser oceanic plate

Convection current in magma

Rock melting back into magma or partial re-melting to form part of the outer mantle

Magma rising up through continental crust

In 1914 Alfred Wegener put forward a theory that the continents were drifting apart. Some of the evidence he used included:

- The continents seemed to fit together like a jigsaw.
- Fossils in Africa matched those in South America.
- Greenland was getting further away from Europe.

Wegener's ideas were not accepted by the majority of scientists until new evidence came to light in the 1960s from the study of the formation of new rocks at the sea floor either side of where a crack occurred. This and subsequent research has backed up his theory, which developed into the theory of plate tectonics. The theory has now gained acceptance by the scientific community.

C2 | Construction Materials

Materials from Rocks

Many construction materials come from substances found in the Earth's crust:

- The metals **iron** and **aluminium** are extracted from rocks called ores.
- Clay is a rock that makes **brick** when it is baked.
- **Glass** is made from sand, which is small grains of rock.

Some rocks, like limestone, marble and granite, just need to be shaped to be ready to use as a building material. **Aggregates** (crushed rock or gravel) are used in road making and in building. Limestone is the easiest to shape because it is the softest; marble is harder to shape and granite is harder still.

Rock is dug out of the ground in mines and quarries. Mining and quarrying companies have to take steps to reduce their impact on the local area and environment because mines and quarries can:

- be noisy
- be dusty
- take up land
- change the shape of the landscape
- increase the local road traffic.

A responsible company will also ensure it reconstructs, covers up and restores any area it has worked on.

Limestone and marble are both forms of calcium carbonate ($CaCO_3$). When calcium carbonate is heated it breaks up into calcium oxide and carbon dioxide.

calcium carbonate → calcium oxide + carbon dioxide
$CaCO_3$ → CaO + CO_2

This type of reaction is called a **thermal decomposition**; one material breaks down into two new substances when it is heated.

When clay and limestone are heated together, **cement** is made. One of the main uses of cement is to make **concrete**.

Concrete

To make concrete you need:

- 1 bag of cement
- 5 bags of sand
- 5 bags of aggregate
- water.

1. Mix all dry ingredients.
2. Add water.
3. Mix well.
4. Allow to set.

Concrete is very hard but not very strong. It can be strengthened by allowing it to set around steel rods to reinforce it. **Reinforced concrete** is a **composite material**.

Rocks and Composite Materials

Rocks differ in hardness because of the ways in which they were made. Limestone is a **sedimentary** rock. Marble is a **metamorphic** rock made from limestone that has been put under pressure and heated, which makes it harder. The hardest rock, granite, is an **igneous** rock.

A composite material combines the best properties of each material. Reinforced concrete combines the strength and flexibility of the steel bars with the hardness and bulk of the concrete. Reinforced concrete has many more uses than ordinary concrete.

Copper

Copper is a very useful metal because it is an excellent conductor of heat and electricity and it does not corrode. Copper is made by heating naturally occurring copper ore with carbon. The reaction removes oxygen from the ore; this is a **reduction** reaction.

The process uses lots of energy, which makes it expensive. It is cheaper to recycle copper than to extract it from its ore. Recycling also conserves our limited supply of copper ore and uses less energy. However, there are some problems with recycling copper. It has to be separated from other metals or there may be metals mixed with it and it is not always easy to persuade the public to recycle materials.

Electrolysis

New and recycled copper is purified by the process of **electrolysis**. The diagram below shows the apparatus used to purify copper:

HT **Electrolysis** is the name given to a chemical reaction that uses electricity. The electricity is passed through a liquid or a solution called an **electrolyte**, e.g. copper(II) sulfate solution is used to purify copper. **Electrodes** are used to connect to the electrolyte. The positive electrode (**anode**) is made of impure copper and the negative electrode (**cathode**) is made of pure copper.

The cathode increases in mass because pure copper is deposited on it from the electrolyte.

$$Cu^{2+} + 2e^- \longrightarrow Cu$$

This is a reduction reaction because the copper ion has gained electrons.

The anode loses mass as the copper dissolves into the electrolyte.

$$Cu - 2e^- \longrightarrow Cu^{2+}$$

This is an oxidation reaction because the copper atom has lost electrons.

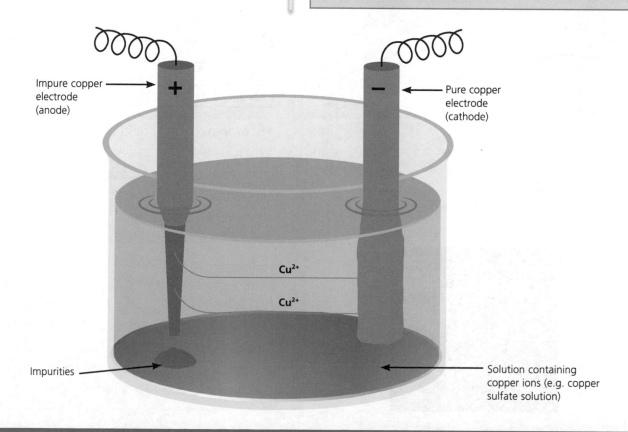

Impure copper electrode (anode)

+

Pure copper electrode (cathode)

−

Cu^{2+}

Cu^{2+}

Impurities

Solution containing copper ions (e.g. copper sulfate solution)

Alloys

An **alloy** is a mixture of a metal with another element (usually another metal). Alloys, e.g. bronze and steel, are made to improve the properties of a metal and to make them more useful – they are often harder and stronger than the pure metal:

- **Amalgam** is made using mercury and is used for fillings in teeth.

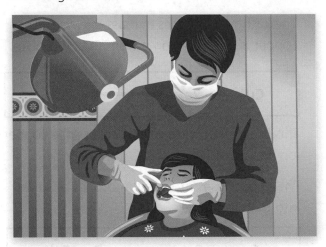

- **Brass** (made of copper and zinc) is used in coins, musical instruments, door handles and door knockers.

- **Solder** (made of lead and tin) is used to join electrical wires.

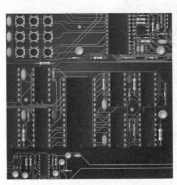

Smart Alloys

A **smart alloy** such as **nitinol** (an alloy of nickel and titanium) is used to make spectacle frames because it can be bent and twisted but it will return to its original shape when it is heated. It has **shape memory**.

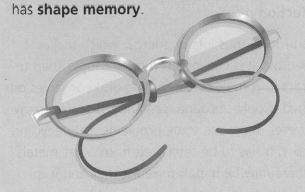

Metals and Properties

You may be asked to compare metals and alloys from a table such as this.

	Order of Hardness	Density (g/cm³)	Melting Point (°C)	Order of Strength
Copper	4	8.9	1083	4
Brass	2	8.6	920	2
Steel	1	7.8	1420	1
Lead	5	11.3	327	5
Solder	3	9.6	170	3

For example, you should be able to work out that lead is denser than solder but that solder is stronger than lead. You may be asked to suggest which metal properties are important for a particular use, e.g. the metal wire inside a vacuum cleaner cable should conduct electricity well and be flexible.

You may be asked to explain why a metal is used for a particular purpose given data such as that in the table. For example, giving more than one reason why brass could be used to make the cables holding up a suspension bridge, or why steel might be a better choice.

Rusting Conditions

The diagram opposite shows an investigation into what conditions are needed to make a nail rust. Four nails were placed in test tubes in different conditions and left for a week. The only nail that rusted was the one in test tube 3. From this we can tell that rusting needs iron, water and oxygen (in air). The addition of oxygen to the iron is an **oxidation reaction**.

Rust flakes off the iron exposing more metal to corrode. Rusting happens even faster when the water is salty or is made from acid rain. Car bodies can rust. They are usually scrapped when this happens because it makes the metal weaker. Aluminium does not rust or corrode in air and water. Instead, it quickly forms a layer of aluminium oxide when it comes into contact with air. This layer stops any more air or water from coming into contact with the metal. This built-in protection will not flake off.

Rusting is an example of an oxidation reaction. This is a reaction where oxygen is added to a substance to make an oxide. Oxygen is added to the iron in the presence of water.

iron + oxygen + water ➡ hydrated iron(III) oxide

Only iron and steel rust; other metals corrode. In your exam you may be asked to interpret information about the rate of corrosion of different metals. For example, the table below gives descriptions of metals in different conditions. You could state that aluminium is the least corroded metal because even in salty wet air, its appearance is the least changed (it has only dulled).

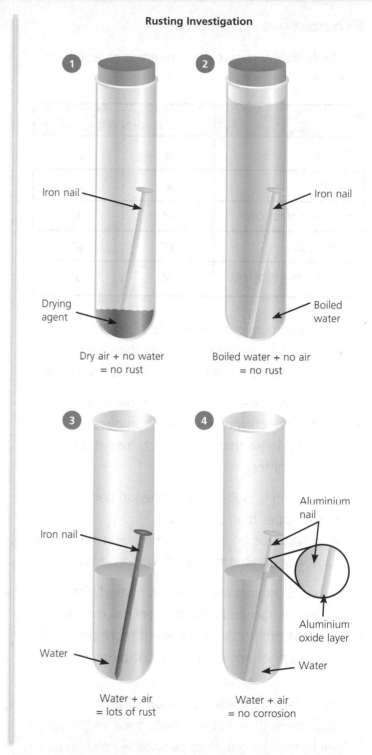

Rusting Investigation

1. Iron nail / Drying agent — Dry air + no water = no rust
2. Iron nail / Boiled water — Boiled water + no air = no rust
3. Iron nail / Water — Water + air = lots of rust
4. Aluminium nail / Aluminium oxide layer / Water — Water + air = no corrosion

Metal	Clean and Dry Air	Wet Air	Acidic Wet Air	Salty Wet Air
Steel	Shiny	Dull (rusty)	Very dull (rusty)	Very dull (rusty)
Copper	Shiny	Dull	Green layer	Green layer
Aluminium	Shiny	Shiny	Dull	Dull
Silver	Shiny	Shiny	Tarnished	Tarnished

Properties of Metals

The table below shows the properties of aluminium and iron:

Property	Aluminium	Iron
Dense	✗	✔
Magnetic	✗	✔
Resists corrosion	✔	✗
Malleable	✔	✔
Conducts electricity	✔	✔

Iron or aluminium can be used to build cars as they can be pressed into shape; they are both malleable. They are both good electrical conductors.

Aluminium is well-suited to the job because:
- it does not corrode (whereas iron does)
- it is less dense than iron (which means the car will be lighter).

However, iron is well-suited to the job because:
- it is cheaper than aluminium
- it is magnetic (aluminium is not) which means it can be separated for recycling more easily.

However, most cars are made from **steel**. Steel is an alloy of iron and carbon. Steel has different properties from iron which make it more useful. It is harder and does not corrode as fast as iron.

> HT Some cars are made from aluminium. Aluminium does not rust or corrode so the car will last for longer. And because aluminium is less dense than steel, the car will:
> - be lighter
> - get better fuel economy.

Aluminium can be mixed with other metals such as copper and magnesium to create an alloy.

Materials in a Car

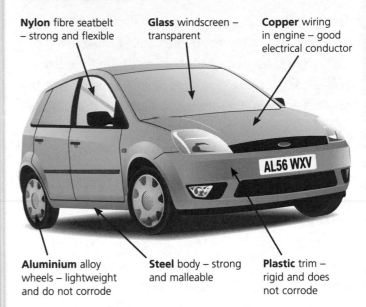

Nylon fibre seatbelt – strong and flexible

Glass windscreen – transparent

Copper wiring in engine – good electrical conductor

Aluminium alloy wheels – lightweight and do not corrode

Steel body – strong and malleable

Plastic trim – rigid and does not corrode

You should be able to suggest what properties are important in a material used in a car, such as the examples in the diagram. If you are given information about the properties of these materials, you should be able to explain why they are used for a particular job in the car.

> HT You may be asked make judgements on the suitability of materials used for car manufacture given all the relevant information.

Recycling

Most materials used in a car can be recycled. From 2006, the law requires that 85% of a car must be able to be recycled; this will increase to 95% in 2015.

The problem is separating all the different materials from each other. However, there are many benefits. Recycling materials means:
- fewer disposal problems
- less energy is needed for extracting them from ores
- limited natural resources will last longer.

Recycling the plastics and fibres reduces the amount of crude oil needed to make them, and conserves oil reserves. There are a number of materials in a car, e.g. lead in the car battery, which would cause pollution if put into landfill, so recycling also protects the environment.

Manufacturing Chemicals: Making Ammonia

Ammonia

Ammonia (NH_3) is an alkaline gas. It is made from nitrogen and hydrogen. Getting these gases to combine chemically and stay combined is very difficult. This is because the reaction is reversible: as well as nitrogen and hydrogen combining to form ammonia, the ammonia decomposes in the same conditions to form hydrogen and nitrogen.

Reversible reactions have the symbol ⇌ in their equation to show that the reaction can take place in either direction.

Ammonia can be used to make nitric acid and fertilisers. Farmers rely on cheap fertilisers made from ammonia to produce enough food for an ever-growing world population.

The Haber Process

Fritz Haber was the first to work out how to make ammonia on a large scale. The raw materials are **nitrogen** (obtained from the air) and **hydrogen** (from natural gas or the cracking of crude oil).

nitrogen	+	hydrogen	⇌	ammonia
N_2	+	$3H_2$	⇌	$2NH_3$

The mixture of 1 part nitrogen and 3 parts hydrogen is compressed to a high pressure and passed into a reactor. The gases are passed over an iron catalyst at 450°C. This is where the reaction takes place (see diagram below).

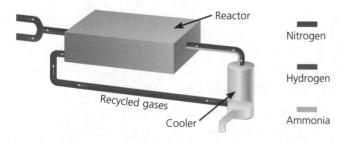

About 28% of the gases are converted into ammonia, which is separated from the unreacted hydrogen and nitrogen by cooling, and collected as a liquid. The unreacted gases are recycled.

You may be asked to interpret graphs and tables containing data about the conditions in the Haber process.

Example

Interpret the graph and table below to explain how temperature and pressure affect the rate of reaction in the Haber process.

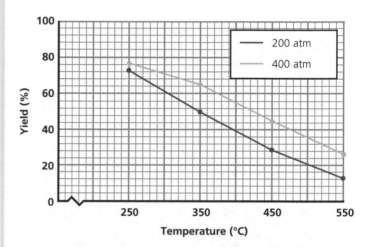

% Yield	Temperature			
Pressure	250°C	350°C	450°C	550°C
200 atm	73%	50%	28%	13%
400 atm	77%	65%	45%	26%

From the above information you should be able to pick out, for example, that the yield falls when temperature is increased and that the yield increases as pressure increases.

> **HT** You may also be asked to interpret data on other industrial processes in terms of rate, percentage yield and cost.

Cost

The cost of making a new substance depends on:
*   the price of energy (gas and electricity)
*   labour costs (wages for employees)
*   how quickly the new substance can be made
*   the cost of starting materials (reactants)
*   the cost of equipment needed (plant and machinery).

Factors Affecting Cost

There are various factors that affect the cost of making a new substance, including:

- the pressure required – the higher the pressure the higher the plant cost
- the temperature required – the higher the temperature the higher the energy cost
- the catalysts required – catalysts reduce costs because they increase the rate of reaction, but they need to be purchased in the first place which increases initial costs
- the number of people required to operate machinery – automation reduces the wage bill
- the amount of unreacted material that can be recycled – recycling reduces costs.

Economic Considerations

Economic considerations determine the conditions used in the manufacture of chemicals:

- The **rate of reaction** must be high enough to produce a sufficient daily yield of product.
- **Percentage yield** achieved must be high enough to produce a sufficient daily yield of product (a low percentage yield is acceptable providing the reaction can be repeated many times with **recycled starting materials**).
- **Optimum conditions** should be used to give the most economical reaction (this could mean producing a slower reaction or a lower percentage yield at a lower cost).

Economics of the Haber Process

$$N_2 + 3H_2 \rightleftharpoons 2NH_3$$

There is great economic importance attached to producing the maximum amount of ammonia in the shortest possible time at a reasonable cost. This demands some compromise.

Effect of Temperature

If a high temperature is used in the Haber process it has two effects:

- It speeds up the rate of the reaction and ammonia is made faster.
- It reduces the percentage yield of ammonia.

The compromise is between making more ammonia slowly at a low temperature and making less ammonia more quickly at a higher temperature.

Effect of Pressure

Using a high pressure in the Haber process gives a higher percentage yield but it is expensive to construct the reaction chamber, and to maintain it to contain such high pressure. The compromise here is between the cost of high pressure and the percentage yield of ammonia.

Effect of a Catalyst

Using a catalyst increases the rate of the reaction, although it does not affect the percentage yield. However, although using a catalyst can reduce costs, purchasing it increases initial costs.

In summary:

- a low temperature increases yield but the reaction is too slow
- a high pressure increases yield but the reaction is too expensive
- a catalyst increases the rate of reaction but does not change the percentage yield.

Therefore a compromise is reached in the Haber process and the following conditions are used:

- An iron catalyst.
- A high pressure of 200 atmospheres.
- A temperature of 450°C which gives a fast reaction with sufficiently high percentage yield.

In your exam, you may be asked to use the above ideas to interpret how rate, cost and yield affect other industrial processes.

Acids and Bases

Acids are substances that contain hydrogen ions (H⁺) in solution. pH is a measure of the concentration of H⁺ ions in the solution. (Acids have a pH of less than 7.) **Bases** are the oxides and hydroxides of metals. The bases that are soluble are called **alkalis**.

The pH of a solution can be determined by using **universal indicator**. You just need to add a few drops of the solution to the substance and compare the resulting colour to the pH chart as shown below.

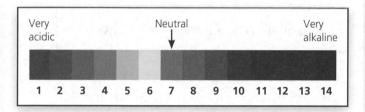

Universal indicator changes colour gradually as the pH changes, whereas some indicators only have one colour change e.g. red-coloured litmus changes suddenly to blue when an alkali is added.

Neutralisation

Acids and bases (alkalis) are chemical opposites. If they are added together in the correct amounts they can cancel each other out. This is called **neutralisation** because the solution that remains has a neutral pH of 7.

For example, adding hydrochloric acid (HCl) to potassium hydroxide (KOH) is a neutralisation reaction:

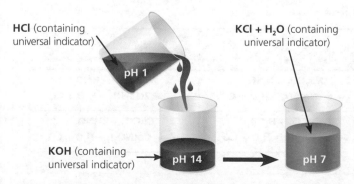

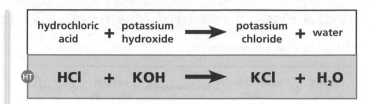

Carbonates **neutralise** acids to produce a salt and water, but they also produce carbon dioxide gas.

Word Equations and Naming Salts

You may be asked to write a word equation for a neutralisation reaction and you may need to work out the name of the **salt** that is produced.

The first name of a salt made by neutralisation comes from the first name of the base or carbonate used, for example:
- **sodium** hydroxide will make a **sodium** salt
- **copper** oxide will make a **copper** salt
- **calcium** carbonate will make a **calcium** salt
- **ammonia** will make an **ammonium** salt.

The second name of the salt comes from the acid used, for example:
- hydro**chlor**ic acid will produce a **chlor**ide salt
- **sulf**uric acid will produce a **sulf**ate salt
- **nitr**ic acid will produce a **nitr**ate salt.
- **phosph**oric acid will produce a **phosph**ate salt.

For example, adding **potassium** hydroxide to **nitr**ic acid to neutralise it will make **potassium nitrate**.

Fit the names of the reactants and products into the general equation. The word equation for this last reaction is:

Word Equations and Naming Salts (cont)

Other examples include:

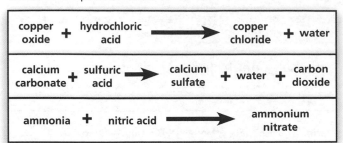

Do not try to use formulae or mix formulae with words if you are asked to write a word equation. Always use + signs and the arrow.

Neutralisation Reactions

Neutralisation can be summarised by looking at what happens to the ions in the solutions:

- Alkalis in solution contain **hydroxide ions**, **OH⁻**(aq).
- Acids in solution contain **hydrogen ions**, **H⁺**(aq).

Neutralisation can therefore be described using the following ionic equation:

$$H^+(aq) + OH^-(aq) \rightleftharpoons H_2O(l)$$

You should be able to construct any of the following balanced formula equations for producing salts:

acid + base → salt + water

	Hydrochloric acid (HCl)	Sulfuric acid (H_2SO_4)	Nitric acid (HNO_3)
Sodium hydroxide (NaOH)	NaOH + HCl → NaCl + H_2O	2NaOH + H_2SO_4 → Na_2SO_4 + $2H_2O$	NaOH + HNO_3 → $NaNO_3$ + H_2O
Potassium hydroxide (KOH)	KOH + HCl → KCl + H_2O	2KOH + H_2SO_4 → K_2SO_4 + $2H_2O$	KOH + HNO_3 → KNO_3 + H_2O
Copper(II) oxide (CuO)	CuO + 2HCl → $CuCl_2$ + H_2O	CuO + H_2SO_4 → $CuSO_4$ + H_2O	CuO + $2HNO_3$ → $Cu(NO_3)_2$ + H_2O

acid + base → salt

	Hydrochloric acid (HCl)	Sulfuric acid (H_2SO_4)	Nitric acid (HNO_3)
Ammonia (NH_3)	NH_3 + HCl → NH_4Cl	$2NH_3$ + H_2SO_4 → $(NH_4)_2SO_4$	NH_3 + HNO_3 → NH_4NO_3

acid + carbonate → salt + water + carbon dioxide

	Hydrochloric acid (HCl)	Sulfuric acid (H_2SO_4)	Nitric acid (HNO_3)
Sodium carbonate (Na_2CO_3)	Na_2CO_3 + 2HCl → 2NaCl + H_2O + CO_2	Na_2CO_3 + H_2SO_4 → Na_2SO_4 + H_2O + CO_2	Na_2CO_3 + $2HNO_3$ → $2NaNO_3$ + H_2O + CO_2
Calcium carbonate ($CaCO_3$)	$CaCO_3$ + 2HCl → $CaCl_2$ + H_2O + CO_2	$CaCO_3$ + H_2SO_4 → $CaSO_4$ + H_2O + CO_2	$CaCO_3$ + $2HNO_3$ → $Ca(NO_3)_2$ + H_2O + CO_2

Fertilisers

Fertilisers are chemicals that farmers use in order to provide their plants with the **essential elements** they need for growth. They increase the crop **yield**. The three main essential elements found in fertilisers are:

- nitrogen, N
- phosphorus, P
- potassium, K.

Fertilisers must be soluble in water so that they can be taken in by the roots of plants.

The following fertilisers can be made by neutralising an acid with an alkali:

- Ammonium nitrate is manufactured by neutralising nitric acid with ammonia.
- Ammonium sulfate is manufactured by neutralising sulfuric acid with ammonia.
- Ammonium phosphate is manufactured by neutralising phosphoric acid with ammonia.
- Potassium nitrate is manufactured by neutralising nitric acid with potassium hydroxide.

Urea made from ammonia can also be used as a fertiliser.

Making a Fertiliser

To make a fertiliser (e.g. potassium nitrate), follow these steps (see diagrams below). Make sure you recognise the apparatus.

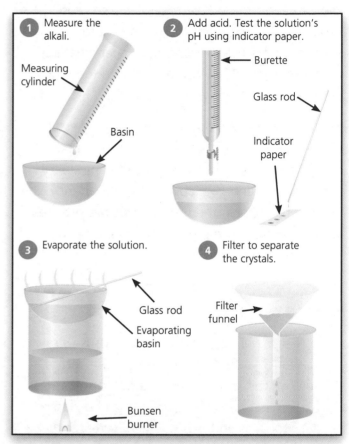

1 Measure the alkali.

Measuring cylinder

Basin

2 Add acid. Test the solution's pH using indicator paper.

Burette

Glass rod

Indicator paper

3 Evaporate the solution.

Glass rod

Evaporating basin

Bunsen burner

4 Filter to separate the crystals.

Filter funnel

HT

1 Measure out the alkali (e.g. potassium hydroxide) into a basin using a measuring cylinder.

2 Add the acid (e.g. nitric acid) from a burette. Use a glass rod to put a drop of solution onto indicator paper to test the pH. Continue to add the acid a bit at a time until the solution is neutral (pH 7).

3 Evaporate the solution slowly until crystals form on the end of a cold glass rod placed in the solution. Leave to cool and crystallise.

4 Filter to separate the crystals from the solution.

5 Remove the crystals, wash them and leave to dry.

This method is another example of producing a salt (a fertiliser) by neutralisation.

Benefits and Problems of Fertilisers

The use of chemical fertilisers helps us grow more food. The world's population is rising, increasing the demand for food. Fertiliser use also causes problems. Chemical fertilisers can pollute water supplies and cause the death of water creatures (**eutrophication**) when they are used too much and are washed into rivers and lakes.

Fertilisers increase crop yield by replacing essential elements in the soil that have been used up by a previous crop, or by increasing the amount of essential elements available. More importantly, they provide nitrogen in the form of **soluble nitrates** which are used by the plant to make protein for growth.

HT ## Eutrophication

Careless overuse of fertilisers can cause stretches of nearby water to become stagnant very quickly. This process is called **eutrophication**:

1. Fertilisers used by farmers may be washed into lakes and rivers, which increases the level of nitrates and phosphates in the water and increases the growth of simple algae.

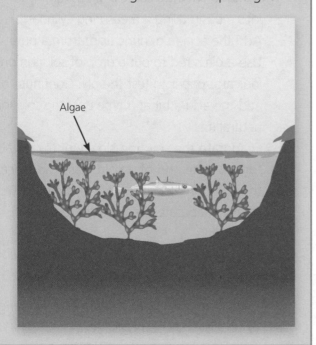

Algae

2. The algal bloom blocks off sunlight to other plants, causing them to die and rot.

Rotting plants

3. This leads to a massive increase in the number of aerobic bacteria (which feed on dead organisms), which quickly use up the oxygen. Eventually, nearly all the oxygen is removed. This means there is not enough left to support the larger organisms, such as fish and other aquatic animals, causing them to suffocate.

Sodium Chloride

Sodium chloride is normal table salt and is used as a flavouring and preservative. It is also a very important raw material for the chemical industry.

In the UK, sodium chloride can be obtained from sea water. It is mined in Cheshire as a solid (rock salt) for gritting roads and also by solution mining for the chemical industry. Salt mining has led to subsidence of the ground in some parts of Cheshire.

The electrolysis of concentrated **sodium chloride solution** (also known as salt solution or brine) forms hydrogen at the **cathode** (negative electrode) and chlorine at the **anode** (positive electrode). Sodium hydroxide is also formed in the solution.

The electrodes must be made from **inert** materials as the products are very reactive.

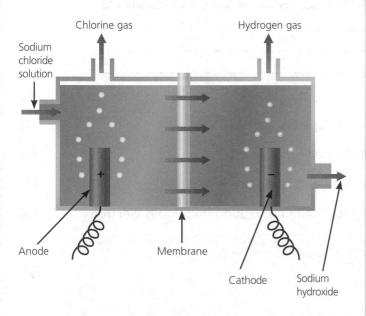

There are many uses of the products obtained from the electrolysis of sodium chloride:
* Hydrogen is used to make margarine.
* Sodium hydroxide is used to make soap.
* Chlorine is used to sterilise water, make solvents and make plastics, for example, PVC.
* Chlorine and sodium hydroxide are reacted together to make household bleach.

The test for chlorine is that it bleaches moist litmus paper.

Test for Chlorine

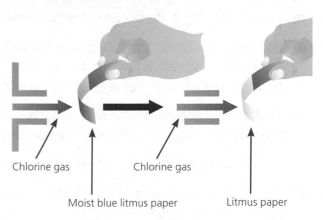

Chlorine gas   Chlorine gas

Moist blue litmus paper   Litmus paper

HT Electrolysis of Sodium Chloride Solution

Sodium chloride solution contains the ions Na^+, H^+, Cl^- and OH^-. The H^+ and Cl^- ions are discharged at the electrodes.

Half-equations can be written for the reactions that take place at the electrodes in the electrolysis of concentrated sodium chloride solution.

* Hydrogen is made at the cathode:

$$2H^+ + 2e^- \longrightarrow H_2$$

Electrons are gained – this is reduction.

* Chlorine is made at the anode:

$$2Cl^- - 2e \longrightarrow Cl_2$$

Electrons are lost – this is oxidation.

Sodium ions and hydroxide ions are not discharged and remain to make sodium hydroxide solution.

The electrolysis of sodium chloride is a very important part of the chemical industry; many other parts of industry depend on the products manufactured by this process.

1 a) What is the lithosphere? [1]

b) Explain why the plate tectonic theory is now accepted by most scientists. [2]

2 a) i) A proposal for a limestone quarry has been submitted and a large number of residents of a nearby village are against it. Explain why. [3]

ii) Some residents do not oppose the quarry. Suggest a reason why. [1]

b) The thermal decomposition of limestone is an important step in the manufacture of cement. Magnesium carbonate also thermally decomposes and makes magnesium oxide and carbon dioxide.

Write a word equation for the thermal decomposition of magnesium carbonate. [1]

3 a) Steel is an alloy composed mainly of iron. Explain why steel is more useful for making support cables in bridges than iron. [2]

b) Name the metal that is mixed with lead to make solder. [1]

c) Give two properties that make solder useful for welding metal gas pipes together. [2]

4 Look at the data table. It shows how temperature and pressure affect the yield (%) of the product in an industrial process.

Pressure (atmospheres)	Temperature (°C)			
	250	350	450	550
50	60%	30%	11%	4%
100	67%	34%	16%	7%
200	73%	50%	29%	14%

a) What temperature gives the lowest yield? [1]

b) What is the yield at 100 atmospheres pressure and a temperature of 350°C? [1]

c) What happens to the yield as the temperature is decreased? [1]

5 Explain why ammonium phosphate, $(NH_4)_3PO_4$, can be used as a fertiliser. [3]

6 a) Name the three important products made when sodium chloride solution undergoes electrolysis. [3]

HT b) Complete and balance the ionic equation for the reaction at the anode during the electrolysis of sodium chloride solution. [2]

$$Cl^- - \underline{\quad} \longrightarrow Cl_2$$

7 Describe what happens when a continental plate and an oceanic plate collide. [3]

8 Describe the conditions used in the Haber process and explain why a temperature of 800°C and a pressure of 50atm would not be suitable. [6]

✏ *The quality of written communication will be assessed in your answer to this question.*

9 A wetland nature reserve has banned the use of chemical fertilisers within a three mile radius. Explain how this protects the water creatures that live in the reserve. [3]

10 Nitric acid (HNO_3) reacts with sodium carbonate (Na_2CO_3) to make sodium nitrate ($NaNO_3$), water (H_2O) and carbon dioxide (CO_2). Write a balanced symbol equation for this reaction. [2]

P1: Energy for the Home

This module looks at:

- How heat and temperature are different, and the use of water to heat homes.
- Insulation, energy efficiency and energy transfer through conduction, convection and radiation.
- Electromagnetic waves and their uses.
- Light and its uses in digital communication.
- Radiations in the electromagnetic spectrum and the properties, dangers and uses of infrared and microwave radiation.
- Infrared radiation and its uses in the home and in transmitting signals.
- Global communication and the benefits, uses and impacts on society of wireless transmission.
- Waves, how they carry information and how they can be harmful to organisms, as well as how climate is affected by natural and human activity.

Heat Flow

Every year we spend millions of pounds heating our houses but much of this heat energy escapes through our windows and roofs. We can use our understanding of temperature and heat flow to help us to reduce our energy usage and save us money.

Temperature

Temperature is a measure of how hot something is. The unit of measurement is **degrees Celsius, °C. Heat** is a form of **energy** and is measured in **joules, J**.

> **HT** **Temperature** is a measurement of how **hot** something is using a chosen scale, usually degrees Celsius, °C, but sometimes degrees Fahrenheit, °F.
>
> Heat is a measurement of **energy** on an **absolute scale** – always in **joules, J**.

If there is a difference in temperature between an object and its surroundings then this results in the flow of heat energy from the **hotter** region to the **cooler** region, making the hotter region cool down and the cooler region warm up. A hot region is the **source**; the cold region is the **sink**.

If an object's **temperature rises** it is **taking in heat energy**. For example, if you take a can of cola out of the fridge it will soon warm up to room temperature because the can and the liquid take in heat energy from the air in the room.

If an object's **temperature falls** it is **giving out heat energy**. For example, a hot cup of tea will soon cool down, eventually reaching room temperature. If you hold it in your hands you will feel the heat energy flowing from the cup into your hands.

When an object has a **very high temperature** compared to its surroundings it will **cool down very quickly**. As its temperature gets nearer room temperature, it will cool down at a slower rate.

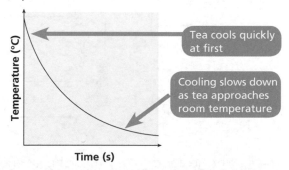

Tea cools quickly at first

Cooling slows down as tea approaches room temperature

> **HT** The molecules in all solid materials are vibrating, so they have **kinetic energy**. The higher the temperature of an object, the higher the average kinetic energy of the molecules.

Thermograms

Temperature can be represented by a range of colours in a **thermogram**.

- The windows are where most heat energy is escaping so they show up as white (hottest), yellow or red.
- The well-insulated loft is where the least heat energy is escaping so this shows up as black (coldest), dark blue or purple.

Measuring Heat Energy

The amount of energy needed to raise the temperature of an object depends on:
- the **mass** of the object
- the **change in temperature** required
- the **specific heat capacity** (see opposite) of the material that the object is made of.

This experiment measures the amount of heat energy needed to change the temperature of an aluminium block.

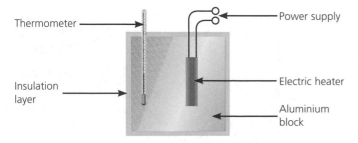

The electric heater provides 100J of heat energy per second. Therefore, if you time how many seconds it takes for the temperature of the aluminium to rise by a certain amount, e.g. 10°C, you can calculate the total amount of energy used to make the temperature rise using the following formula:

| Total energy supplied | = | Energy supplied per second | × | Number of seconds |

It takes 50 seconds to raise the temperature of the aluminium block by 10°C.

So, 100J/s × 50s

= 5000J ← The total energy supplied

Specific Heat Capacity

The **specific heat capacity of a material** is the energy needed to raise the temperature of **1kg** of the material by **1°C**.

Each material has its own value, which measures how much energy it needs to raise its temperature by 1°C. For example, it takes more energy to raise the temperature of a liquid by 1°C than to raise the temperature of a solid by 1°C (liquids have higher specific heat capacities than solids).

The following equation is used to find the amount of energy required to raise the temperature of an object by a certain amount:

| Energy (J) | = | Mass (kg) | × | Specific heat capacity (J/kg/°C) | × | Temperature change (°C) |

Example 1

The specific heat capacity of copper is 387J/kg/°C. Calculate how much heat energy is required to raise the temperature of a 5kg block of copper by 10°C. Use the formula:

$$\text{Energy} = \text{Mass} \times \text{Specific heat capacity} \times \text{Temperature change}$$

$$= 5\text{kg} \times 387\text{J/kg/°C} \times 10\text{°C} = \textbf{19 350J}$$

Example 2

It takes 28 800J of heat energy to raise the temperature of a 4kg block of aluminium from 22°C to 30°C. Calculate the specific heat capacity of aluminium. Rearrange the formula:

$$\text{Specific heat capacity} = \frac{\text{Energy}}{\text{Mass} \times \text{Temperature change}}$$

$$= \frac{28\,800\text{J}}{4\text{kg} \times 8\text{°C}} = \textbf{900J/kg/°C}$$

Melting and Boiling

The data below shows how the temperature of some water in a kettle changed over 300 seconds.

Time (s)	0	30	60	90	120	150
Temperature (°C)	21	39	55	68	79	88
Time (s) (cont)	180	210	240	270	300	
Temperature (°C)	95	100	100	100	100	

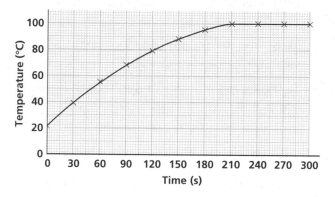

The temperature rises quickly to begin with, but once it gets to 100°C it stops rising – it remains **constant**. The temperature of the water will never rise above **100°C**, no matter how long it is heated for. This is because when the water reaches 100°C, all the **heat energy** supplied by the kettle is being used to **boil** the water instead of heating it up.

Likewise, when you put an **ice cube** in a drink it does not **melt** immediately because it needs **energy** to warm it to 0°C, and then more energy to make it melt – its temperature stays at 0°C while it melts.

The temperature of a material does not change when it is **boiling**, **melting** or **freezing** (i.e. changing state). So, to **interpret data** showing the heating or cooling of an object, look for places where the temperature stays the same.

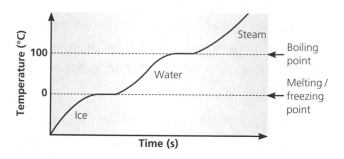

HT During the melting and boiling of water, the energy supplied is used to break **intermolecular bonds** (instead of raising its temperature) as the water changes state from solid to liquid and from liquid to gas. This explains why the temperature of the material does not change – none of the energy is being used to heat up the material. All the energy is being used to break bonds.

Specific Latent Heat

The amount of heat energy required to melt or boil 1kg of a material is called its **specific latent heat**. It depends on:

- the **material**
- the **state** (solid, liquid or gas).

The energy required to boil or melt a certain mass of a material can be found using the equation:

Energy (J) **=** **Mass** (kg) **✗** **Specific latent heat** (J/kg)

Example

The specific latent heat of ice is 330 000J/kg. An ice sculpture at 0°C with a mass of 10kg is left to melt. Calculate the amount of energy required to melt the ice. Use the equation:

Energy = Mass × Specific latent heat

= 10kg × 330 000J/kg

= **3 300 000J**

Explaining Heat Transfer

Conduction and Insulation

Materials that allow heat energy to spread through them quickly are called **conductors**. **Metals** are good conductors of heat.

Materials that allow heat energy to spread through them much more slowly are called **insulators**. Most **non-metals,** such as **wood**, **plastic**, **glass** and **air**, are good insulators.

Examples of conductors and insulators include:

- A saucepan is made of a good conductor to get the heat to the food, e.g. copper or aluminium.
- A saucepan handle is made of a good insulator to stop the heat getting to your hand, e.g. wood or plastic.
- Clothing and bedding are both good insulators, because they trap air within their material, and between the layers to stop heat leaving your body.
- Curtains are good insulators because they trap a layer of air between them and the window, which helps to reduce heat energy loss by conduction.

Convection Currents

When air next to a radiator in a room gets warm it will expand and become less dense, so it will rise up and cooler air will move in to take its place. So hot air rises and warms up the air already at ceiling level. This movement of air is called a **convection current**.

Circulation of Air Caused by a Radiator

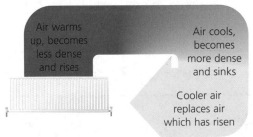

Air warms up, becomes less dense and rises

Air cools, becomes more dense and sinks

Cooler air replaces air which has risen

Reflecting Infrared

Heat can also move as **infrared radiation** (a type of electromagnetic wave). Infrared can be reflected from shiny surfaces but is very easily absorbed by dark or rough surfaces. This is why putting a shiny surface behind a hot object will reflect the heat, but a dark surface will get very hot.

Why Are Things Hot?

If an object is heated, its particles will start to move more quickly – so a hotter object has faster moving particles. In a solid the particles move by vibrating. In a liquid and a gas the particles move around, flowing past each other and moving further apart – which is why hot air is less dense and rises.

Some examples of heat transfer are listed below:

- A hot-water tank is made of stainless steel, which reduces heat loss by radiation.
- Hot water tanks usually have an insulating jacket made of foam to reduce heat loss by conduction and convection.
- Refrigerators are insulated to reduce heat gain by conduction and convection.

Reducing Heat Losses in the Home

Apart from curtains, there are many ways to reduce heat loss from a home. It is important to think about the **payback time** – how long it takes to pay for the **insulation** from the savings you make. The diagram shows how heat can escape from a house, and the methods that can reduce heat loss.

Insulation Method	Cost	Annual Saving	Payback Time
❶ Fibreglass loft insulation	£400	£80	5 years
❷ Reflective foil on walls behind radiators	£40	£10	4 years
❸ Cavity wall insulation	£600	£30	20 years
❹ Double glazing	£1800	£60	30 years
❺ Draught excluders	£40	£40	1 year

Saving Energy in the Home

Houses lose energy through doors, windows, floors, walls and roofs. Energy is lost from a **source** to a **sink**, so when heat is lost from the home, the home is the source and the atmosphere is the sink. Each design feature in the house on the previous page helps to save energy by reducing heat loss. The table below explains how.

Method of Insulation	Reduces...	How?
Fibreglass loft insulation	• Conduction • Convection	• By trapping layers of air (a good insulator) between the fibres.
Reflective foil on walls behind radiators	• Radiation	• By reflecting infrared heat energy into the room.
Cavity wall insulation	• Conduction • Convection	• By trapping air (a good insulator) in the foam.
Double glazing	• Conduction • Convection	• By trapping air (a good insulator) between the panes of glass.
Draught excluders	• Convection	• By keeping as much warm air inside as possible.

HT Each method of house insulation helps to save energy by conduction, convection and radiation.

Other energy-saving strategies include drawing curtains early in the evening to reduce heat loss by convection, having carpets and sealed floors to stop heat loss through the floor and keeping inside doors to conservatories closed in cold weather.

Energy Efficiency

Energy efficiency is a measure of how good an appliance is at converting **input energy** (the energy supplied) into **useful output energy**.

For a television, the input energy is electrical energy and the useful energy output is light and sound. We need to be able to see and hear the programmes. But televisions also produce heat energy which, in this case, is wasted energy.

This equation is used to calculate energy efficiency:

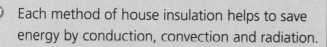

$$\text{Efficiency \%} = \frac{\text{Useful output energy (J)}}{\text{Total energy input (J)}} \times 100\%$$

Example

An old style 60 watt light bulb uses 60 joules of electrical energy every second. In 50 seconds it gives out 300 joules of light energy (useful energy). Use the formula to calculate the efficiency of the light bulb:

$$\text{Efficiency} = \frac{\text{Useful output energy}}{\text{Total energy input}} \times 100\%$$

$$= \frac{300}{60 \times 50} \times 100\%$$

$$= 0.1 \times 100\% = \mathbf{10\%}$$

Sankey diagrams show how much energy is transferred and where it is lost. The Sankey diagram below shows the energy delivered to customers from a power station. By working out the energy used and the energy lost at each stage the actual values can be added to the diagram to show the efficiency.

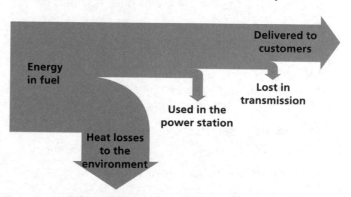

Conduction

Conduction is the transfer of heat energy through a substance from a hotter region to a cooler region without any movement of the substance itself. As a substance, (e.g. a metal poker for a fire) is heated, the kinetic energy of its particles increases (they vibrate more). This kinetic energy is transferred between the particles in the poker and, gradually, energy is transferred along it.

Heat energy is also transferred by the free electrons in a metal. The free electrons flow, collide with and transfer energy to other atoms, increasing their kinetic energy and so the temperature rises. This is why metals are good conductors.

Insulation only stops heat loss by conduction. Heat energy can still be lost from a hot object by **radiation** and/or by **convection**. A cavity wall in a house should be filled with foam to trap air and to prevent heat loss by convection and conduction.

Convection

Because fluids (liquids) and gases can flow, they can transfer heat energy from hotter to cooler regions by their own movement. As the liquid or gas gets hotter, its particles move faster causing it to expand and become less dense. It will then rise up and be replaced by colder, denser liquid or gas.

Radiation

Radiation is the transfer of heat energy by **electromagnetic waves**. Hot objects emit mainly infrared radiation, which can pass through a vacuum, i.e. no material is needed for its transfer. How much infrared radiation is given out or taken in by an object depends on its surface:

- Dark matt surfaces emit more radiation than pale shiny surfaces at the same temperature.
- Dark matt surfaces are better absorbers (poorer reflectors) of radiation than pale shiny surfaces at the same temperature.

Conduction

Convection

Circulation of Air Caused by a Radiator

Air warms up, becomes less dense and rises

Air cools, becomes more dense and sinks

Cooler air replaces air which has risen

Radiation

Dark surface emitting more radiation than light surface

5 mins later

5 mins later

Transverse Waves

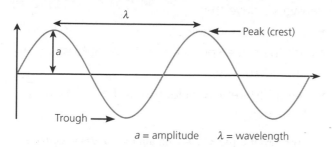

a = amplitude   λ = wavelength

All **transverse waves** have the same features:

- **Amplitude** is the maximum disturbance caused by the wave at a trough or crest (peak).
- **Wavelength** is the distance between corresponding points on two successive disturbances (i.e. from one peak to the next peak).
- **Frequency** is the number of waves produced (or that pass a particular point) in one second.

Transverse Wave

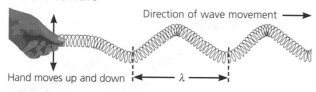

Longitudinal Wave

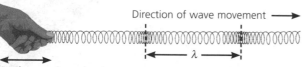

The following formula shows the relationship between wave speed, frequency and wavelength:

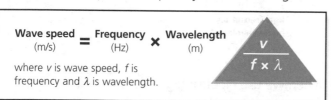

Wave speed (m/s) = **Frequency** (Hz) × **Wavelength** (m)

$$\frac{v}{f \times \lambda}$$

where v is wave speed, f is frequency and λ is wavelength.

Example 1

Calculate the speed of this wave if its frequency is 5Hz.

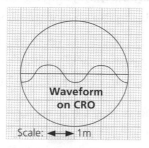

Waveform on CRO

Scale: ◄──► 1m

Wave speed = Frequency × Wavelength

= 5Hz × 2m

= **10m/s**

Example 2

Radio 5 Live transmits on a frequency of 909kHz. If the speed of radio waves is 300 000 000m/s, on what wavelength does it transmit? Rearrange the formula:

$$\text{Wavelength} = \frac{\text{Wave speed}}{\text{Frequency}} = \frac{300\ 000\ 000\text{m/s}}{909\ 000\text{Hz}}$$

= **330m**

Example 3

A radio station transmits on a wavelength of 1500m. What is the frequency of the transmission?

$$\text{Frequency} = \frac{\text{Wave speed}}{\text{Wavelength}}$$

$$= \frac{3 \times 10^8}{1.5 \times 10^3} = \mathbf{2 \times 10^5 Hz}$$

Reflection

Usually, when a ray of light hits a shiny surface it is **reflected**:

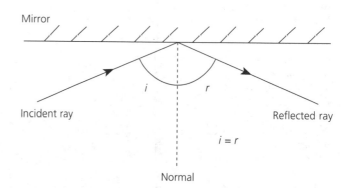

i = r

Multiple Reflections

Light can reflect off a series of surfaces, just like a ball bouncing off the walls in a squash court.

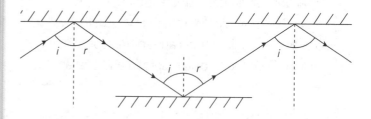

Refraction

Usually, when a ray of light or infrared passes from one material into another it changes direction – it is **refracted**. This happens because the wave changes speed:

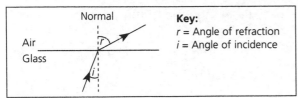

Key:
r = Angle of refraction
i = Angle of incidence

Diffraction

Diffraction is when a wave spreads out because it has passed through a narrow opening about the size of the wavelength of the wave.

Diffraction can also happen if the wave passes an obstacle. If the obstacle is large then the wavelength has to be large, otherwise there is no effect.

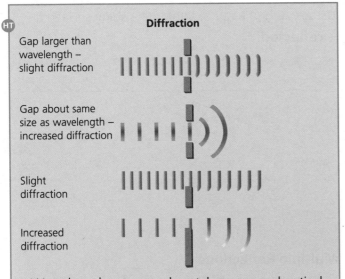

Wave-based sensors such as telescopes and optical microscopes are affected by diffraction because diffraction will limit the smallest thing they can detect. For example, two stars close together will be seen as one single star because diffraction will make the light from the two stars spread and overlap.

The Electromagnetic Spectrum

Light is one part of the **electromagnetic spectrum**. Together with the other forms of electromagnetic radiation, it makes a continuous spectrum that extends beyond each end of the **visible spectrum** (light).

Each type of electromagnetic radiation is a transverse wave that:
- travels in straight lines
- has the same speed through space (a vacuum) – 300 000 000m/s
- has a different wavelength and a different frequency.

The Seven Types of Electromagnetic Waves in Order of Frequency and Wavelength

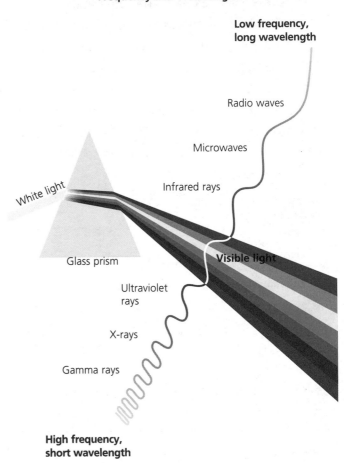

When we use electromagnetic waves for communication, the size of the receiver has to be similar to the size of the wavelength. So, satellite dishes have small receivers at their centre to detect microwaves, while television aerials are much bigger in order to detect radio waves.

Radio waves are used for TV and radio stations. Microwaves are used to communicate using satellites. Infrared and light waves are used to communicate using optical fibres.

Communicating Through Signals

Communicating through electromagnetic waves (including light) is far faster than any other means. **Morse code** was an early communication system. This is a series of 'on-off' signals with each letter of the alphabet being made up of a different pattern of dots and dashes.

P H Y S I C S
.--. .... -.-- ... .. -.-. ...

Morse code was an early form of **digital** communication – digital means a code made up of a pattern of two types of input e.g. 'ons and offs'.

Signals were relayed between stations to increase the distance a message could be sent.

Now we use digital codes to transmit information along **optical fibres** as pulses of light or infrared. Using different colours (wavelengths) of light, it is possible to send many different messages along a fibre at the same time.

Communicating with Light

Like all **electromagnetic waves**, light travels very fast. The reason why modern communication is so fast is because it uses light as a signal, but the signal needs to be sent as a digital code – flashes of light.

An **optical fibre** is a long, flexible, transparent glass or plastic fibre of very small diameter. Light can travel the length of the fibre by reflecting off the sides (total internal reflection). This transfer of light depends on the **critical angle** of the substance.

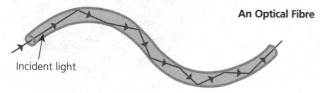

An Optical Fibre

Incident light

Critical Angle

Different media have different critical angles. The critical angle is the maximum angle (measured from the normal) at which light can be refracted and escape from a material.

At angles larger than the critical angle, the light is totally reflected back into the material. The critical angle depends on the refractive index of the medium. A large refractive index gives a smaller critical angle.

Total Internal Reflection

When the angle of incidence is bigger than the **critical angle**, the light or infrared is **totally internally reflected** and not refracted.

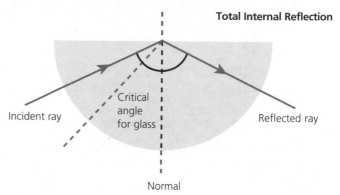

Total Internal Reflection

Incident ray

Critical angle for glass

Reflected ray

Normal

Total internal reflection can only happen inside a material that is more optically dense than the material surrounding it, such as inside glass surrounded by air, or inside Perspex surrounded by air.

Total internal reflection can also happen to light in water when the light approaches the surface and is reflected back into the water instead of emerging into the air. In all these cases the **angle of incidence** of the light has to be bigger than the **critical angle** for total internal reflection to happen. If the angle is too small some light will be refracted and escape into the air.

Comparing Communication Signals

Light, radio and electrical signals can be used for communication – they each have advantages and disadvantages:

- Light is fast but sometimes requires optical fibres.
- Radio waves can travel further, via satellites if necessary, but the signal can be lost.
- Electrical signals are reliable and can be boosted if they get weak, but they require wires.

Lasers

Lasers form very narrow beams of light of a single colour (single wavelength). They have many uses including medical (cutting skin during surgery or during dental treatment which means the patient bleeds less), cutting materials in industry, guiding weapons by reflecting off the target to show that the aim is correct, or in laser light shows.

A Medical Laser

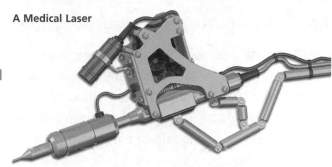

HT Lasers produce an intense beam of light in which all the light waves:

- have the same frequency
- are **in phase** with each other
- do not diverge.

'In phase' means that all the peaks and troughs match up: they go up together and down together. For this to happen the waves must be **monochromatic** and **coherent**. When waves are **in phase** they transmit a lot of energy.

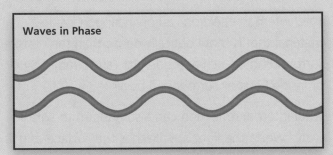

Waves in Phase

Waves **out of phase** will transmit less overall energy.

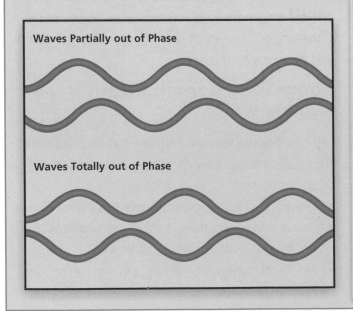

Waves Partially out of Phase

Waves Totally out of Phase

Compact Discs

Digital information can be stored as a sequence of billions of tiny pits (a digital code) in a metal layer on the underside of a compact disc (CD). A CD player spins the disc and laser light is reflected from the pits. The reflected pulses of light (a digital signal of ons and offs) are detected and turned into electrical signals and then transmitted to an amplifier.

Using Signals

Light signals travel very fast and can be sent down optical cables using total internal reflection with only a very small amount of signal loss.

Electrical signals can be sent along wires, but the resistance of the wire causes the signal to deteriorate.

Radio signals can travel through air but are easily lost or weakened in the atmosphere.

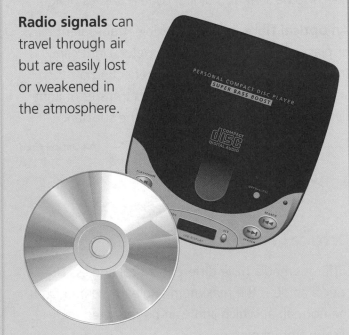

Cooking and Communicating using Waves

P1

Infrared Cooking

Most cooking still uses **infrared radiation**, such as from a toaster, a hob or an oven. All objects give out infrared radiation, but:

- hotter objects emit more infrared than cooler ones
- black objects emit more infrared than pale ones at the same temperature
- rougher or dull surfaces emit more infrared than shiny ones at the same temperature.

Food in an oven gets hot because it absorbs the infrared that the oven is emitting, but:

- a black object absorbs infrared better than a pale one so gets hotter more easily
- a shiny object reflects infrared so it cannot absorb it and does not get hot so easily.

Wrapping a baked potato in kitchen foil reduces the amount of infrared radiation it emits, so it cools more slowly.

> **HT** When an object absorbs infrared, the energy from the electromagnetic wave is absorbed by the particles in the surface of the object and so their kinetic energy increases (they vibrate more). This means that the temperature of the object will increase. Some surface particles become so hot that they get damaged and become brown (burnt).
>
> The hotter surface particles then transfer energy to particles in the centre by conduction so the whole thing cooks through.

Cooking with Infrared

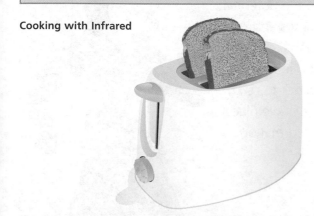

Cooking with Microwaves

Microwaves are electromagnetic waves that can penetrate about 1cm into food and are absorbed by water and fat molecules in the food. Shiny metal surfaces reflect microwaves, which is why you should not put metal objects into a microwave oven. Microwaves can penetrate through glass or plastic so food in containers made from these materials can cook easily.

> **HT** Water and fat molecules in the outer layers of food absorb microwaves very efficiently and the energy increases their kinetic energy, making the material hot. The food should be stirred to enable the heat to be spread through the food.
>
> Because microwaves have longer wavelengths than infrared, they are absorbed by water and fat more easily than infrared. This means they cook food more quickly and are more able to cause burns on body tissue.

Mobile Communicating with Microwaves

Microwaves are used for communication, such as by mobile phones. They can transmit information over large distances that are in **line of sight**, which means that the transmitters and receiver must have no obstacles between them. Some areas are not in line of sight so they have poor signals, which is why your mobile phone may cut out or fail to get a connection.

Microwaves Used for Communication

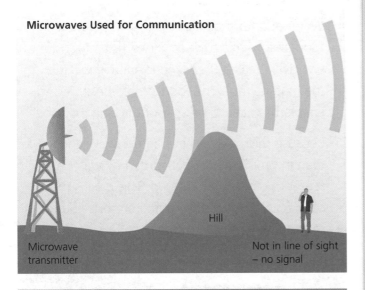

Microwave transmitter

Hill

Not in line of sight – no signal

HT Sometimes microwave signals are poor. This could be due to:
- a large obstacle such as a building or a mountain
- the signal being affected by the weather or a large area of water
- the curvature of the Earth
- the signal getting weaker as it travels further.

It is possible to try to reduce signal loss by putting transmitters and receivers closer together and by positioning them on high ground so that buildings, hills etc. do not get in the way and so that the curvature of the Earth has less effect.

It is not possible to rely on diffraction of microwaves around buildings, hills etc. because their wavelength is not large enough. It is also impossible to do anything to stop interference between signals.

Microwaves and Health

Mobile phones use microwave signals but they are not the same wavelength as those used in microwave cookers.

There is public concern about children using mobile phones because their skulls are thinner and people think their brains may get damaged by the microwaves. There is also concern about possible dangers to adult users and people who live near transmission masts, even though there is no evidence of any effect on people when thorough tests have been carried out.

There have been many studies into the effect of mobile phones on people. Almost all show that there is no adverse effect on people who use mobile phones or live near phone masts. Some studies show a link between frequent phone use and brain tumours but they do not show that the microwaves from phones or phone masts cause these tumours or any other effect on health.

All the scientific studies into the effects of mobile phones' microwave radiation have been published. This makes it possible to check all the results and compare the conclusions.

HT There seems to have been conflicting information about the impact of mobile phones or phone masts on people.

It is always important to weigh up the evidence and the potential benefits and risks and make an informed decision, and not to rely on a single piece of information.

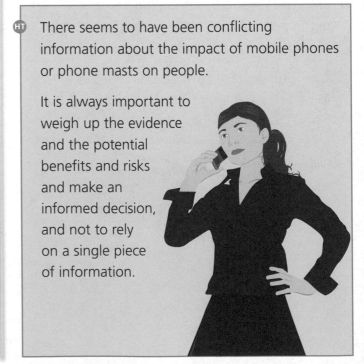

Transmitting

Infrared waves can be used for transmitting signals for:

- remote controls for televisions, videos, DVDs etc.
- short distance links between computers and printers, e.g. WiFi.

Infrared sensors, such as those connected to burglar alarms in houses, can detect body heat. When a warm body walks into a room the burglar alarm will go off.

Thermal imaging cameras also detect infrared from bodies and can display an image to show which parts are warm or cold (see page 78).

Signals

There are two types of signal that can be used to transmit data. They are:

- **analogue**
- **digital**.

They each have properties that make them suitable for different uses.

Analogue signals vary continually in amplitude. They can have any value within a fixed range of values and are very similar to the sound waves of speech or music.

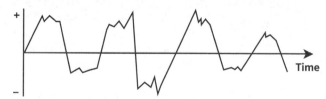

Analogue Signal

Digital signals do not vary; they have only two values or states: on (1) or off (0). There are no values in between. The information is a series of pulses.

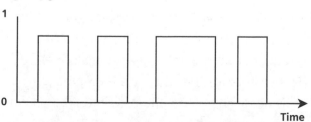

Digital Signal

Remote Controls

When you use a remote control, the device in your hand uses a set of digital signals (or codes) to control each of the different functions.

Advantages of Digital Signals

The big advantage of digital signals is that more information can be transmitted along optical fibres. **Multiplexing** is a technique where two or more digital signals can be carried down the same fibre.

Both digital and analogue signals suffer from interference in the form of noise, but this is easily removable from digital signals leaving them as clear as when they were first sent.

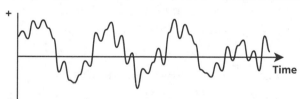

Analogue Signal – poor signal quality due to interference (noise)

Digital Signal – high signal quality because interference is easily removed

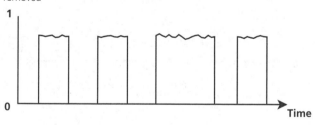

Optical Fibres

Optical fibres have big advantages:

- they allow very rapid transmission of data
- they use light pulses for transmission of data.

The switchover to digital TV from analogue TV means that we will be able to receive more channels because more information can be transmitted by digital signals and images are less likely to be fuzzy or 'noisy' because digital signals do not suffer from interference.

Wireless Signals

Electromagnetic radiation (such as radio waves and microwaves) can be used to send information without optical fibres because it can be reflected and refracted in the same way as visible light. Sometimes this is a disadvantage because it can cause loss of signal.

This **wireless technology** (**WiFi**) is used in **TVs**, **radios**, **mobile phones** and **laptop** computers and has three main advantages:

- No wiring is needed – you do not need to be connected directly to the transmitter.
- It enables items to be portable and convenient.
- It allows access signals on the move.

But you do have to have an aerial to pick up the signals.

HT Transmitting Signals

Long-distance communication depends on:

- **Reflection** – the **ionosphere** is an electrically charged layer in the Earth's upper atmosphere. **Longer wavelength** radio waves are **reflected** by the ionosphere. This enables radio and television programmes to be transmitted between different places, which may be in different countries or continents around the Earth.
- **Refraction** – at the interfaces of different layers of the Earth's atmosphere, refraction of waves results in the waves changing direction.

Refraction and reflection in the ionosphere act in the same way as total internal reflection for light and keep the signals in the Earth's atmosphere instead of escaping into space.

Negative effects on signal quality include:

- **Diffraction** (changes to the direction and intensity of waves) at the edge of transmission dishes causes the waves to spread out, which results in signal loss. Interference from similar signals limits the distance between transmitters. Positioning transmitters in high

places can help to overcome the nuisance of obstacles blocking signals.
- **Refraction** at the interfaces of different layers of the Earth's atmosphere can lead to loss of signal.

If signals are sent to satellites above the atmosphere, the signals can be 'bounced' from satellite to satellite (received and retransmitted) to cover longer distances and travel round the world.

Interference

Some signals have a poorer quality than others so sometimes the information is 'noisy' (see page 89) and you get hissing, called **interference**. Another form of interference can happen if there are two radio stations using similar frequencies. The two stations can cause mutual interference so that you hear parts of one programme on top of the other.

HT

Interference cannot happen with digital signals because each station has a transmission frequency that is well separated from all others (each station requires less bandwidth so there is more space) and the processing enables random signals to be filtered out before they are turned into sound or light.

DAB

Some radio stations provide better signals than others. Analogue signals often suffer from interference. **DAB (Digital Audio Broadcasting)** uses digital signals. The advantages are:

- more stations can be transmitted
- there is less interference between stations.

But there are some disadvantages to DAB:

- The audio quality is poorer than with **FM** (frequency modulated signals – analogue).
- Not all areas of the country are covered so you may be driving along and pass into an area where there is no signal.

Earthquakes

Earthquakes produce shock waves that can travel inside the Earth and across the surface and cause damage to buildings and the Earth's surface. These waves are called **seismic waves** and can be detected by **seismometers** and recorded on **seismographs**. There are two types of seismic wave:

- **P-waves** (primary waves) are longitudinal and travel through both solids and liquids.
- **S-waves** (secondary waves) are transverse waves and travel through solids but not through liquids. They travel more slowly than P-waves.

Earthquakes can also cause tsunamis.

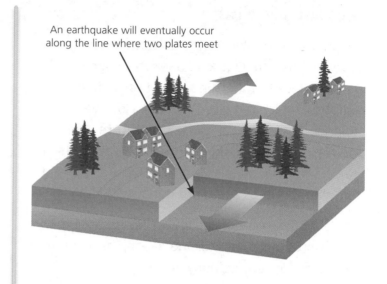

An earthquake will eventually occur along the line where two plates meet

HT The properties of seismic waves provide evidence for the structure of the Earth. After an earthquake occurs, the waves are detected all over the world as shown in the diagram. P-waves can travel through the liquid outer core so they are detected in most places. S-waves will not pass through liquids so they are only detected closer to the **epicentre** (the centre of the earthquake).

Primary Waves (P-waves)

Primary waves are longitudinal waves: the vibrations are in the same direction as the wave is travelling. They can travel through solids and liquids and through all the layers of the Earth.

Secondary Waves (S-waves)

Secondary waves are transverse waves: the vibrations are at right angles to the direction the wave is travelling. They can travel through solids but not liquids. They cannot travel through the Earth's outer (liquid) core and are slower than primary waves.

A study of seismic waves indicates that the Earth is made up of:

- a thin crust
- a mantle which is semi-fluid and extends almost halfway to the centre
- a core which is over half of the Earth's diameter with a liquid outer part and a solid inner part.

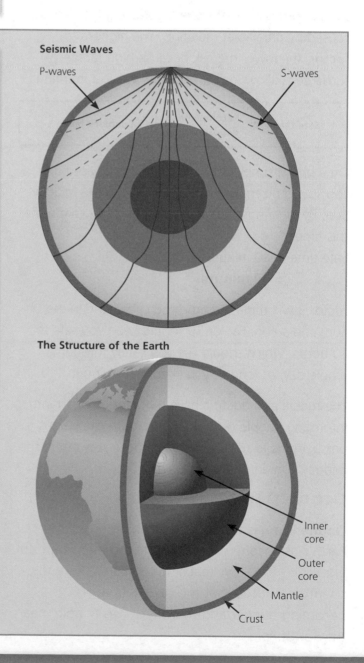

Seismic Waves

P-waves

S-waves

The Structure of the Earth

Inner core

Outer core

Mantle

Crust

Dangerous Sun

The Sun is responsible for life on Earth but it can also be very damaging. One kind of electromagnetic wave produced by the Sun is called **ultraviolet radiation** which can give a sun tan. However, prolonged exposure can cause **premature skin ageing**, **sunburn** or **skin cancer**, and **cataracts** in the eyes of older people. Some people are suffering from skin cancer as a result of using sunbeds.

People with darker skin are less at risk because their skin absorbs more ultraviolet radiation, meaning less reaches the underlying body tissue.

Sunscreen is also effective at reducing the damage caused by ultraviolet radiation. The higher the **factor**, the lower the risk because high factor sunscreens allow longer exposure without burning:

Safe time	=	Recommended exposure time	×	Sunblock factor

On a bright sunny day in England it is advised that you spend no more than 20 minutes in the midday Sun. By wearing a factor 2 sunblock you can double this time:

Safe time = 20 minutes × 2

= **40 minutes**

Factor 3 will triple the time you can spend in the Sun, and so on. Factor 30 should keep you safe for 10 hours, which should see you safely through a whole day, as long as you keep reapplying it.

The media and doctors have put a lot of effort into informing people of the dangers, with campaigns to persuade us to use sunscreens, not to stay out in the midday Sun and not to use sunbeds.

Ozone Depletion

Ozone is a gas found naturally high up in the Earth's atmosphere – it prevents too many harmful ultraviolet (UV) rays reaching the Earth. Recently, scientists noticed that the **ozone layer** is becoming thinner and that there was a 'hole' near the South Pole.

To be certain that this was a real effect and not simply an error, they have repeated the readings with new equipment and other scientists have also taken the same measurements. Scientists have also made predictions, based on their explanation of how the ozone hole was formed, and then tested their predictions to show that their explanations are reliable.

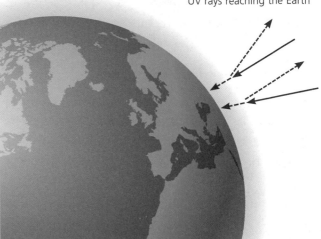

Ozone layer prevents some UV rays reaching the Earth

HT Scientists think that **CFCs** (chlorofluorocarbons) in factories, fridges and aerosol cans have caused this change in the ozone layer. They have tested this idea by looking at how ozone is affected by different chemicals and also to see for how long CFCs remain in the upper atmosphere. As a result of their work CFCs have been banned in most countries. Scientists are continuing to measure the ozone levels, which do vary every year. The prediction is that the ozone layer will eventually repair itself.

Scientists are also noting that more people are suffering from skin cancer in the Southern Hemisphere. This fits with the observation that the ozone layer is important in protecting people from ultraviolet radiation. There have been many campaigns to educate people to use sunscreens. People are now more likely to use sunscreen and less likely to get sunburnt.

1. Sarah is eating an ice lolly in the Sun and the ice lolly starts to melt. The graph shows the change in temperature of the ice lolly over time.

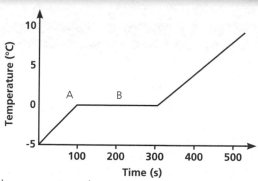

a) What is happening at point B? Put a tick (✓) next to the correct option. **[1]**

Sarah has finished the ice lolly ☐   Sarah has gone inside, out of the sun ☐

The ice lolly is melting ☐   The ice lolly is freezing ☐

b) Specific heat capacity is the energy needed to raise the temperature of 1kg of a material by 1°C. The mass of the ice lolly is 57g and the specific heat capacity is 1.34 kJ/kg/°C. Use this information, and the graph above to calculate the energy transferred to the ice lolly before point A. **[1]**

2. Helen wants to insulate her house. She has collected some information about different types of house insulation, as shown in the table.

Method of Insulation	Cost	Annual Saving	Payback Time
Double glazing	£2100	£70	
Draught excluders	£30	£20	
Fibreglass loft insulation	£500	£25	

a) Fill in the table by calculating the payback time for each method of insulation. **[3]**

b) Use the information in the table to suggest which type of insulation Helen should choose. Explain your answer. **[1]**

c) Because her house has been so cold recently, Helen has been putting silver foil on the walls behind the radiators. How might this have helped to keep the house warm? **[3]**

3. Mobile phones are an example of the use of wireless technology.

a) What type of electromagnetic wave do mobile phones use? **[1]**

b) A mobile phone signal has a wavelength of 2cm. What is the frequency of the wave? **[2]**

c) There are plans for the construction of a new mobile phone mast in a town due to poor mobile phone signals in the area. The picture shows the location of the town. Identify the likely reason why there is bad signal and explain why this would cause the problem. **[2]**

d) There have been a number of complaints from local residents about the mobile phone mast. State one advantage and one disadvantage of this new construction. **[2]**

The Sun

The Sun is a stable source of energy that transfers energy to the Earth in the form of light and heat.

Photocells

A **photocell** (sometimes called a solar cell) has a flat surface that captures as much of the light energy from the Sun as possible.

It transfers this light energy into an electric current that travels in the same direction all the time (**direct current, DC**).

The power output of the photocells depends on the surface area exposed to the sunlight, so lots of photocells can be joined together to create a larger surface and therefore increase the amount of light captured from the Sun.

The advantages and disadvantages of photocells are listed in the table.

Advantages
- Require little maintenance once installed.
- Can operate in remote locations to give access to electricity without installing power cables.
- No need for fuel because the Sun is the source of energy.
- Have a long life.
- Use energy from the Sun, which is a renewable energy source.
- No pollution or waste produced.

Disadvantages
- Expensive to buy.
- No power at night or during bad weather.

ⓗ How Photocells Work

The Sun's energy is absorbed by the photocell, causing **electrons** to be knocked loose from the silicon crystals. These electrons flow freely as an electric current. This flow of charge is called an **electric current**.

The power of a photocell depends on the surface area exposed to the light, the distance from the light source and the **intensity** of that source. (Intensity means how concentrated the light energy is.)

To maximise power output, an efficient solar collector must track (follow) the Sun in the sky. This requires additional technology, which increases the initial set-up cost.

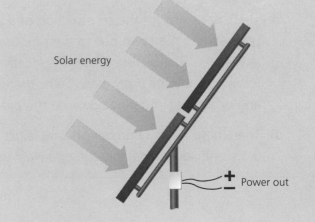

Solar energy

Power out

Harnessing the Sun's Energy

Light from the Sun (radiation) can be captured and used in other ways too.

Sunlight can be absorbed by the surface of a solar panel and transferred into heat energy. Water passed over this surface will be heated to a reasonable temperature and can be used for washing and for heating buildings.

Flat plate collector – solar panel filled with water

Hot water to house

Cold water supply

Heat exchanger

Water tank

Pump

A **curved mirror** can be used to focus the Sun's light, rather like a magnifying glass, to heat up a solar panel.

Glass can be used to provide **passive solar heating** for buildings. 'Passive solar heating' simply means a device that traps energy from the Sun (e.g. a greenhouse) but that does not distribute the energy or change it into another form of energy. This is what causes conservatories to get so hot in the summer.

HT Glass is transparent to some wavelengths of the Sun's radiation, including visible light – so light passes straight through glass windows and is absorbed by objects in the room. The heated surfaces then emit infrared radiation (longer wavelength than visible light), which is reflected back into the room by the glass so cannot escape. This warms up the room.

Wind Turbines

The Sun heats up the air and causes convection currents – the wind. **Wind turbines** depend on the wind produced by the Sun's energy. Wind turbines transfer the **kinetic energy** (**KE**) of the air into **electrical energy**.

The advantages and disadvantages of wind turbines are listed below.

Advantages
• Wind is a **renewable** energy source so it will not run out.
• There is no chemical **pollution** or **waste**.

Disadvantages
• Wind turbines require a large amount of **space** in a windy region to deliver a reasonable amount of electricity.
• They are **dependent** on the wind – no wind means no electricity.
• They cause **visual pollution** because they are very big.

The Dynamo Effect

Electricity can be generated by moving a wire, or a coil of wire, near a magnet (or by moving the magnet near to the wire). When this happens the wire cuts through the lines of force of the magnetic field and a current is produced in the wire, providing it is part of a **complete circuit**.

Electricity can be generated by moving a magnet towards a coil of wire.

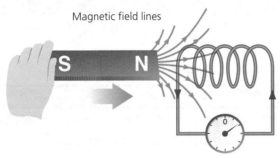

Magnetic field lines

Electricity can also be generated by moving the coil of wire towards the magnet.

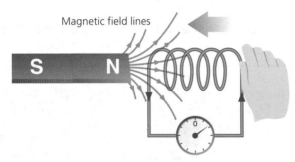

Magnetic field lines

The current generated can be increased by:
- using a stronger magnet
- using more turns of wire in the coil
- moving the coil (or magnet) faster.

Generators use the dynamo effect to produce electricity. **Batteries** produce a **direct current** (DC) but generators produce an **alternating current** (AC). This means that the direction of the current is continually alternating (changing direction) as time passes, every time the magnet (or coil) changes direction.

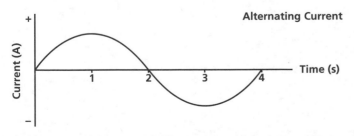

Alternating Current

The graph shows that as time passes, the line curves up and down again. This means that the current alternates from a positive direction (forward) to a negative one (backward) and back again.

The **voltage** generated changes direction and this causes the current to change direction giving alternating current (AC).

N.B. It is also possible to plot the voltage of AC against time.

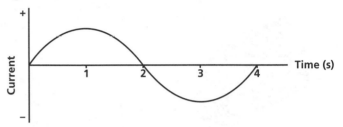

How an AC Generator Works

In an AC generator, a coil of wire is rotated in a magnetic field. In practice, the coil and magnetic field should be close together. As the coil cuts through the magnetic field a current is generated in the coil. The current alternates, i.e. it reverses direction of flow every half turn of the coil, as can be seen below.

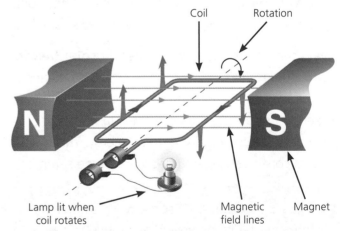

Coil   Rotation

Lamp lit when coil rotates   Magnetic field lines   Magnet

Position of Coil to Generate Current in the AC Cycle

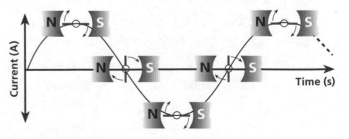

Producing Electricity

The heat energy for our power stations comes from a variety of energy sources. Electricity is transmitted to consumers (homes, factories, offices, farms, etc.) through the network of cables called the **National Grid**, which connects the power stations to consumers.

Fossil fuels such as crude oil, coal and natural gas can be burned to release heat energy, which boils water to produce steam. The steam rotates the turbines, which turn the generators, which produce electricity.

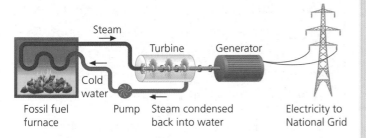

In a **nuclear power station** there are fuel rods of uranium (or sometimes plutonium). They release energy (fission) as heat to make steam in the power station.

Biomass such as wood, straw and manure, can be burned or else fermented to generate methane (a gas) that can be burned in a power station.

Biomass such as wood absorbs carbon dioxide as it grows and then releases it when burned so the amount of CO_2 in the atmosphere remains constant. However, biomass requires very large areas of land to produce enough material to generate significant amounts of energy. This land could be used to grow food instead.

Fossil fuels are readily available now but supplies are running out. They also produce **pollution** including carbon dioxide (CO_2) which is a **greenhouse gas** and contributes to climate change.

Efficiency of a Power Station

A significant amount of energy produced by conventional power stations is wasted. At each stage in the electricity generation process, energy is transferred to the surroundings in a 'non-useful' form, usually as heat.

Below is a typical energy transfer diagram for the process that shows how much energy is wasted at each stage. Only 30J is used usefully; 70J is wasted (also see page 81).

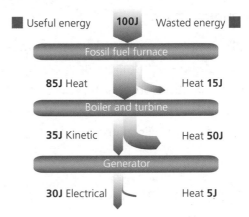

The efficiency of a power station can be worked out as:

$$\text{Efficiency} = \frac{\text{Useful energy output}}{\text{Total energy input}} \times 100\%$$

HT The following equations are also used when calculating the efficiency of a power station:

$$\text{Fuel energy input (J)} = \text{Waste energy output (J)} + \text{Electrical energy output (J)}$$

$$\text{Efficiency} = \frac{\text{Electrical energy output (J)}}{\text{Fuel energy input (J)}} \times 100\%$$

Example

A power station uses 200 000J of fuel energy to produce 80 000J of electrical energy.

a) What is the waste energy output?

Rearrange the first formula:

$$\text{Waste energy output} = \text{Fuel energy input} - \text{Electrical energy output}$$

$$= 200\,000J - 80\,000J$$

$$= \textbf{120\,000J}$$

b) What is the efficiency of this power station?

$$\text{Efficiency} = \frac{\text{Electrical energy output}}{\text{Fuel energy input}} \times 100\%$$

$$= \frac{80\,000J}{200\,000J} \times 100\%$$

$$= 0.4 \times 100\% = \textbf{40\%}$$

The Greenhouse Effect

There are some gases in the Earth's atmosphere that do not allow heat to escape (radiate) into space. This causes the Earth to warm up in the same way that the glass in a greenhouse makes the greenhouse warm up on a sunny day. This warming up is called the **greenhouse effect**.

The greenhouse effect happens because some wavelengths of the electromagnetic spectrum cannot pass through the atmosphere – in particular, longer wavelength infrared rays are absorbed by some gases in the atmosphere – while all the rest can.

Greenhouse Gases

Some examples of greenhouse gases are:

- carbon dioxide
- water vapour
- methane.

All of these gases occur naturally in the atmosphere but are also man-made (i.e. produced through the activities of humans).

Water vapour

Carbon dioxide

How Does the Atmosphere Heat Up?

Short wavelength radiation from the Sun, such as visible light, infrared and ultraviolet, can pass through the atmosphere. This radiation is then absorbed by Earth and heats it.

The Earth is cooler than the Sun so it emits longer wavelength infrared rays that cannot pass through the atmosphere. Instead they are absorbed by the greenhouse gases in the atmosphere, which then heats up.

Methane (produced by cows and by growing rice) is a very powerful greenhouse gas – it absorbs four times as much heat as carbon dioxide. However, carbon dioxide levels are rising much faster than water vapour or methane levels because carbon dioxide is produced by the burning of fossil fuels for heat or to generate electricity, so carbon dioxide is considered to be the biggest cause of global warming.

In examinations you may be expected to interpret data about the abundance and impact of greenhouse gases.

Climate Change

The world is using much more energy than ever before. To generate this energy we need to burn more fossil fuels, and this produces more greenhouse gases such as carbon dioxide. This may be why the Earth is warming up.

Trees can absorb carbon dioxide but humans are cutting down trees (**deforestation**) so the carbon dioxide levels just keep rising and the Earth continues to warm up.

As the Earth warms up, the wind and weather patterns across the globe change – climate change.

Changing weather patterns mean there is more **rainfall** and **flooding** in some countries, more **hurricanes** in other countries and less rainfall and more **droughts** in others (it is not just about everywhere getting warmer!). This is called **climate change**.

Reasons for Climate Change

Human activity and natural phenomena seem to affect the climate:

- Humans burning fossil fuels causes an increase in the greenhouse gases in the atmosphere.
- Dust from factories can reflect heat from cities back to the Earth.
- Dust from volcanic ash clouds can reflect radiation from the Sun back into space and cause the Earth's temperature to fall.

Evidence for Climate Change

Scientists have been measuring the Earth's temperature over many centuries, but it is not easy. Although we can use thermometers today, no one used thermometers to measure the Earth's temperature hundreds of years ago. So scientists have to look at other evidence for the Earth's temperature many years ago, such as the thickness of tree rings in old timbers – if the Earth was warm in a particular year the trees grew more and so the tree ring is thicker.

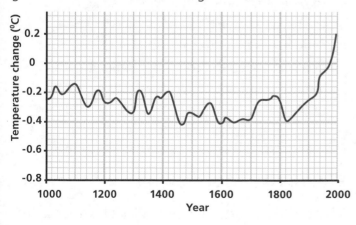

Scientists all over the world are measuring the temperature in thousands of different locations to try to decide if the average temperature is still changing. Of course each different location gives a different answer so it's important that scientists share their data with each other.

The evidence so far seems to show that the Earth has warmed up significantly over the past 200 years. Scientists consider that this trend will continue and predict that the Earth's temperature will continue to increase.

Some people do not believe the evidence because they notice that the local weather is not improving. They mix up the evidence with their own opinion. They forget that scientists use thousands of thermometers and take averages across the world, whereas these people just see the effect on one tiny part of the world.

Scientists in Disagreement

The aim of every scientist is to look critically at other scientists' work to see if it is correct and to point out when it is not. This is how errors in science are found and removed and how science moves forward. If scientists did not do this then science would be far less reliable.

Scientists always take data seriously and will look at every possible way that the data can be interpreted to be sure that they all agree on the correct interpretation. Scientists certainly do not 'take another person's word for it'. Climate change has made scientists work harder than ever to be sure that the data are correct and that the interpretation is agreed.

So far it seems that the data do show that the Earth is warming up and experiments show clearly that greenhouse gases contribute to this. This is where scientists agree.

Some scientists have pointed out that there may be other interpretations of the data on what causes the temperature to rise – could it be Sun spots? Or something that happens on a thousand-year cycle? There is still work to be done to find out if human activity is the main cause of the temperature rising.

In the exam you may be expected to interpret data about increased global warming and climate change as a result of human activity.

Power and Energy Transfer in Circuits

An **electric current** involves a flow of electric charge which transfers energy from the battery or power supply to the components in the circuit. If the component is a light bulb then some of the electrical energy is transformed into light.

The rate of this energy transfer determines the power of the component or device and is measured in joules per second or **watts** (W), where 1 watt is the transfer of 1 joule of energy in 1 second.

Electrical power is calculated using the relationship:

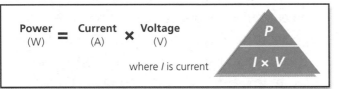

Example

Calculate the power of a lamp when the current flowing through it is 0.3A and the voltage across it is 3V. Using the formula:

Power = Current × Voltage

$$= 0.3A \times 3V$$

$$= \mathbf{0.9W}$$

This means that the lamp transfers 0.9 joules of electrical energy in every second that it is switched on. The power rating of a device is usually written on it in watts (W) or in kilowatts (kW). One kW is 1000W.

HT Example

Calculate the current flowing through a 900W iron when it is being used at its maximum power and working voltage (230V). Rearrange the formula:

$$\text{Current} = \frac{\text{Power}}{\text{Voltage}}$$

$$= \frac{900W}{230V}$$

$$= \mathbf{3.9A}$$

HT Off-Peak Electricity

Electricity is supplied to our homes 24 hours a day. It is often our preferred source of energy because no smoke or gases are produced in the home. However, the electricity is generated using fossil fuels that do produce pollution so the consumer is still, indirectly, adding to the damage being caused to the environment.

Most electricity is used when people are awake and active. There is obviously less demand during the night. However, it is not easy to switch a power station off so power stations generate electricity all the time. (Electricity is produced as you use it – it cannot be stored). In order to encourage people to use the electricity generated during this quiet period, some electricity-generating companies offer cheaper electricity at night called **off-peak electricity**. This can be used for:

- heating up water and storage heaters
- powering washing machines and dishwashers running at night.

The advantage for electricity-generating companies is that they can sell the electricity they generate and do not have to switch off the power stations. The advantage for the consumer is that the cost is lower. The disadvantage to the consumer is that it is inconvenient to run appliances at night because of the noise. The disadvantage to the producer is that they make less profit.

Electricity Meters

Your electricity meter at home will look similar to the one shown. It will show a count of **Units**. These Units represent **kilowatt hours** (kWh), which are a measure of how much electrical energy has been used.

Kilowatt Hours

The number of Units of electricity used by an appliance depends on:

- the power rating in kilowatts of the appliance
- the time in hours that the appliance is switched on for.

To calculate the **cost** of using a device, use the formulae:

Number of kilowatt-hours (kWh) (Energy supplied)	=	Power rating (kW)	×	Time (h)

Total cost	=	Number of kilowatt-hours (kWh)	×	Cost of one kWh

Example

A 1.5kW electric hot plate was switched on for 2 hours. How much does the electricity used cost if electricity is 8p per kWh?

First, calculate the number of kilowatt hours used, using the formula:

$$\text{Number of kWh} = \text{Power} \times \text{Time}$$
$$= 1.5\text{kW} \times 2\text{h} = \textbf{3kWh}$$

Then, calculate the cost:

$$\text{Total cost} = \text{Number of kWh used} \times \text{Cost per Unit}$$
$$= 3 \times 8\text{p} = \textbf{24p}$$

Measuring Energy Supplied

The kilowatt hour is a measure of how much electricity has been used. It is also a measure of how much electrical energy has been supplied.

Energy supplied (kWh)	=	Power (kW)	×	Time (h)

Example 1

A 200W CD player is used for 90 minutes. Calculate the energy supplied.

Using the formula:

$$\text{Energy supplied} = \text{Power} \times \text{Time}$$
$$= 0.2\text{kW} \times 1.5\text{h}$$
$$= \textbf{0.3kWh}$$

Power must be in kW and time must be in h.

Example 2

On a building site, 2.25kWh of electrical energy is supplied to an electric drill in 3 hours. What is the power rating of the electric drill?

Rearrange the formula:

$$\text{Power} = \frac{\text{Energy supplied}}{\text{Time}}$$
$$= \frac{2.25\text{kWh}}{3\text{h}}$$
$$= \textbf{0.75kW or 750W}$$

N.B. To do these calculations, you must always remember to make sure the power is in kilowatts and the time is in hours.

Transformers

To supply electricity to our homes efficiently across the **National Grid** the current is fed through **transformers**.

A transformer can increase the voltage of the supply (and at the same time reduce the current). A transformer can also reduce the voltage of a supply.

Increasing the voltage has the effect of reducing the energy lost as heat as the electricity is transmitted and so reduces the costs.

Using a transformer to increase the voltage of the supply will result in the current decreasing. If the current through the cables is lower then there is less heating ($P = I^2R$). This means that there is less energy wasted as heat in the cable and so the process of transmission is more efficient and less costly.

Comparing Fuel and Energy Sources

The table below lists the advantages and disadvantages of different types of fuel and renewable energy sources. Fossil fuels, biomass and nuclear fuels are commonly used in power stations.

Source	Advantages	Disadvantages
Fossil fuel, e.g. coal, oil, gas	• Enough reserves for short to medium term. • Relatively cheap and easy to obtain. • Coal-, oil- and gas-fired power stations are flexible in meeting demand and have a relatively short start-up time. • Burning gas produces very little SO_2 (sulfur dioxide).	• Produces CO_2 that causes global warming and SO_2 (except burning gas) that causes acid rain. • Removing SO_2 from waste gases to reduce acid rain, and removing CO_2 to reduce global warming, adds to the cost. • Oil is often carried between continents in tankers, leading to risk of spillage and pollution. • Expensive pipelines and networks are often required to transport oil and gas to the point of use.
Biomass, e.g. wood, straw, manure	• It is renewable. • Can be burned to produce heat. • Fermenting biomass produces methane, which can be burned.	• Produces CO_2 which damages the environment. • Large area needed to grow trees which could be used for other purposes, e.g. growing food.
Nuclear fuel, e.g. uranium	• Cost and rate of fuel use is relatively low. • Can be situated in sparsely populated areas. • Nuclear power stations are flexible in meeting demand. • Does not produce CO_2 or SO_2. • High stocks of nuclear fuel. • Can reduce the use of fossil fuels.	• Radioactive material can stay dangerously radioactive for thousands of years and can be harmful. • Storing radioactive waste is very expensive. • Building and decommissioning nuclear power stations are costly. • Comparatively long start-up time. • Radioactive material could be emitted. • Pollution from fuel processing. • High maintenance costs.
Renewable sources, e.g. wind, tidal, hydroelectric, solar	• No fuel costs during operation. • No chemical pollution. • Often low labour costs. • Do not contribute to global warming or produce acid rain. • Produce clean electricity. • Can be constructed in remote areas.	• With the exception of hydroelectric, they produce small amounts of electricity. • Take up lots of space and are unsightly. • Unreliable (apart from hydroelectric), depend on the weather and cannot guarantee supply on demand. • High initial costs to build and install. • High maintenance costs.

Types of Radiation

Radioactive materials are substances that give out nuclear radiation all the time, regardless of what is done to them.

Nuclear radiation is often used in medical treatments such as **radiotherapy** to cure cancer. This works because the nuclear radiation can kill the cancer cells. However, the radiotherapists have to be careful as the nuclear radiation can also kill healthy cells.

Radioactivity involves a change in the structure of the radioactive atom and the release of one of the three types of nuclear radiation:

- **alpha** (α)
- **beta** (β)
- **gamma** (γ).

Nuclear radiation can cause **ionisation**, producing positively and negatively charged ions when atoms lose or gain electrons. Ionisation can be harmful inside the body – it damages cells and can initiate chemical reactions by breaking molecular bonds.

Use of Alpha Radiation

Most **smoke detectors** contain americium-241, which emits alpha radiation. The emitted alpha particles cause the **ionisation** of air particles. The positive and negative ions formed are attracted to oppositely charged electrodes in the detector. This results in a current flowing in the circuit.

When smoke enters the space between the two electrodes, less ionisation takes place because the alpha particles are absorbed by the smoke particles. A smaller current then flows, and the alarm sounds.

Uses of Beta Radiation

A **tracer** is a small amount of a radioactive substance that is put into a system so that its progress through the system can be followed using a radiation detector. A beta-emitter tracer can be used for the following:

- To detect tumours in certain parts of a patient's body, e.g. brain, lungs.
- To identify plants that have been fed with a fertiliser containing a beta particle emitter. (This method can be used to develop better fertilisers.)
- In a **paper thickness gauge**. When beta radiation passes through paper, some of it is absorbed. The greater the thickness of the paper, the greater the absorption. If the paper thickness is too great, then more beta radiation is absorbed, and less passes through to the detector. A signal is then sent to the rollers to move closer together, which reduces the thickness of the paper.

Uses of Gamma Radiation

Gamma radiation can be used to **treat cancer** because it destroys cancerous cells. A high, calculated dose is aimed at the tumour (cancer) from many different angles so that only cancerous cells are destroyed, and not healthy cells.

Gamma radiation can also be used to **sterilise medical equipment** because it can destroy microorganisms – for example, bacteria. An advantage of this method is that no heat is required, which minimises the damage to equipment that heat might cause.

Non-destructive tests can be carried out on welds using gamma radiation. A gamma source is placed on one side, and any cracks or defects can be identified using a detector (e.g. photographic film) on the other side as the gamma rays get through.

Radiation and Cancer

Radiation, including radiation from radioactive waste from **hospitals** and **power stations**, can affect living cells and can cause cancer. It is not a pollutant and does not contribute to global warming.

Dealing with Radioactive Waste Materials

Some radioactive waste can be reprocessed, but often it has to be disposed of. Low-level waste is sealed and buried in landfill sites. High-level waste is mixed with sugar, bonded with glass, poured into a steel cylinder and kept underground.

How to Test Penetrating Power

To test the **penetrating power** of alpha, beta and gamma radiation, a **Geiger counter** can be set up to detect the radiation. Each radioactive source is placed so that its radioactive decay particles can be detected by the Geiger counter. Different materials are then inserted, as in the diagram below.

The diagram below shows each type of radiation's penetrative power, i.e. what materials the radiation can pass through.

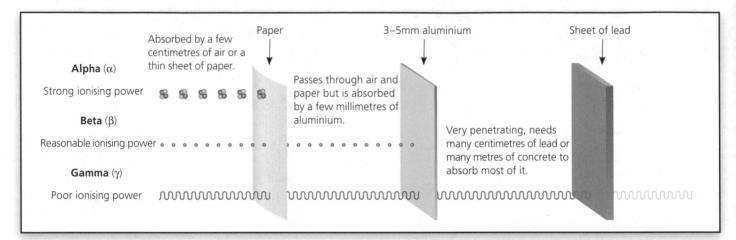

Handling Radioactive Materials Safely

There are four main **safety measures** that need to be taken by people who handle radioactive materials:

- Protective clothing must be worn.
- Tongs should be used to hold the material whenever possible.
- Exposure time should be minimised.
- Radioactive materials must be stored in clearly labelled, shielded containers so that others are aware of what they are handling.

Nuclear Power and Radiation

Nuclear power stations use **uranium** fuel rods to release energy as heat. Uranium, like coal, is a non-renewable resource so it will run out one day, but it does not cause global warming because it does not release carbon dioxide.

One of the waste products from nuclear reactors is **plutonium**, which can be used to make **nuclear bombs.**

HT Advantages and Disadvantages of Nuclear Power

Advantage
Does not emit polluting gases such as carbon dioxide so does not contribute to the greenhouse effect.

Disadvantage
Waste is radioactive so has to be treated and stored for many years.

Problems of Dealing with Radioactive Waste

Some of the problems of dealing with radioactive waste include:

- The waste can remain radioactive for many years.
- Terrorists may try to use it to build dirty bombs.
- It has to be stored carefully to ensure that it does not enter the water supply.
- Acceptable radioactivity levels may change over time so measures may need modifying.

The Universe

The Universe consists of:

- **planets, comets** and **meteors**
- **stars** – our Sun is a star. Stars can be clearly seen, even though they are far away, because they are very hot and give out light
- **galaxies**, which are large groups of stars
- **black holes**, which are dense, dead stars with a very strong gravitational field.

The Solar System

The **Solar System** is made up of the Sun (which is in the centre) surrounded by planets, comets and satellites. The Moon is a natural satellite that orbits the Earth.

Relative Sizes

Planets vary in size but they are all bigger than **meteors** (shooting stars).

Comets are bigger than meteors and have a core of frozen gas and dust. They are up to 20km in diameter. A comet's tail is formed by the core evaporating in the sunlight and always points away from the Sun.

> **HT** The eight planets move around the Sun in paths called **orbits**, which are slightly squashed circles (**ellipses**).
>
> The planets, comets and satellites are kept in their orbits by the **gravitational force** of the larger body they are orbiting. Gravity provides the centripetal force needed to prevent the planets from flying off.

Elliptical Orbits

The planets and satellites travel in circular (or near circular) paths around a larger object. Comets travel in elliptical orbits.

They stay in their orbits because the larger object exerts an **inward pull force** on them. This inward pull force is due to **gravity** and is called the **centripetal force**, e.g. the Earth orbits the Sun because of the Sun's gravitational force.

The Sun and the Planets in the Solar System

Sun

Mercury

Venus

Earth

Mars

Jupiter

Saturn

Uranus

Neptune

Pluto

Not to scale

Pluto has been known as a dwarf planet since September 2006.

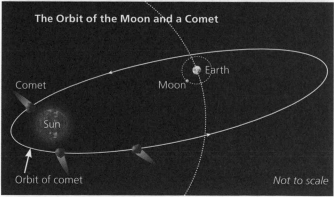

The Orbit of the Moon and a Comet

Comet

Earth

Moon

Sun

Orbit of comet

Not to scale

Relative Sizes (cont)

Stars are much bigger and (except the Sun) are outside of our Solar System. The nearest is *Proxima centauri* which is approximately 4.2 light years away from us.

A **light year** is the distance that light travels in a year. The light year is a useful unit for measuring astronomical distances.

Manned Space Travel

Space is a very dangerous place. There are many difficulties facing a **manned space mission** to the planets:

- The planets are very, very far away so it takes months or even years to reach them.
- The fuel required takes up most of the spacecraft.
- Room must be found to store enough food, water and oxygen for the whole journey.
- A stable artificial atmosphere must be maintained in the spacecraft.
- The temperature in space is about –270°C, so keeping warm is vitally important.
- Outside of the Earth's magnetic field, humans need shielding from cosmic rays.
- The low gravity affects people's health, making muscles weak.
- Radio signals carrying messages take a very long time to reach home.

Unmanned Space Travel

A far more realistic option is to explore our Solar System using **unmanned spaceships**. As well as being able to withstand conditions that are lethal to humans, unmanned probes would not require food, water or oxygen. Once they arrive, probes could be used to send back information about the planet's:

- temperature
- magnetic field
- radiation levels
- gravity
- atmosphere
- surrounding landscape.

However, they cannot bring back samples for analysis.

HT Once the probe arrives on a planet, it can send information to Earth using radio waves which travel at the **speed of light**.

There are many advantages and disadvantages of using **unmanned** spacecraft to explore the Solar System. For example, lower costs – no need to provide food and other things for passengers, and there is also no problem with safety.

However, **reliability** has to be high because there will be no-one to fix any breakdowns, and instruments must require **zero maintenance**.

The Earth and Moon

The Moon may be formed from the remains of a planet that collided with the Earth billions of years ago. If this happened then:

- the planets collided
- the iron cores merged to form the core of the Earth
- the less dense material flew off and now orbits the Earth – it is our Moon.

> **HT** Evidence exists that the Moon may be formed from the remains of a planet that collided with Earth. The Moon is made mainly of rock with no iron core, and this rock appears to have been baked, which would indicate heat generated in a collision.

Asteroids

Asteroids are rocks left over from the formation of the Solar System. They normally orbit the Sun in a belt between Mars and Jupiter, but occasionally they get knocked off course and head towards Earth.

When an asteroid collides with the Earth there can be several devastating consequences:

- The impact forms a crater, which could trigger the ejection of **hot rocks**.
- The heat may cause widespread **fires**.
- Sunlight could be blocked out by the **dust** from the explosion.
- **Climate** change.
- Whole species could become **extinct**, which could, in turn, affect other species.

There is good evidence to suggest that asteroids have collided with the Earth many times in the past:

- **Craters** can be found all over the planet.
- There are layers of unusual **elements** found in rocks.
- There are sudden changes in the number of **fossils** found in adjacent rock layers, which could be due to the sudden death of many animals.

> **HT** There is an asteroid belt between Jupiter and Mars. The asteroids (small lumps of rock) could combine to form a planet but Jupiter's strong gravitational pull prevents them from doing this.

Comets

A **comet** is a small body with a core of frozen gas and dust – like a dirty snowball – which comes from the objects orbiting the Sun far beyond the planets. Their characteristic tails are a trail of debris.

Comets have highly elliptical orbits around the Sun. The speed of the comet increases as it approaches the Sun due to the pull of the Sun's gravity.

> **HT** The comet's speed increases as a result of the increase in the strength of gravity as it approaches the star that it is orbiting. It can also be affected by the gravity of planets.

Near Earth Objects (NEOs)

A **Near Earth Object** (NEO) is an asteroid or comet on a possible **collision course** with Earth. **Telescopes** are used to observe these objects in an attempt to determine their **trajectories** (probable paths). It is difficult to observe NEOs because they move very quickly and there are so many of them (and many artificial satellites in orbit around the Earth).

> **HT** NEOs may pose a threat to the human race but there are actions we can take to reduce that threat. We can:
> - **survey** the skies with telescopes to identify likely NEOs as early as possible
> - **monitor** their progress with satellites
> - **deflect** the object with an explosion if a collision does seem likely.

The Big Bang Theory

One theory that has been used to explain the evolution of the Universe to its present state is the **Big Bang** theory, which states that the whole Universe started billions of years ago in one place with a huge explosion, i.e. a big bang.

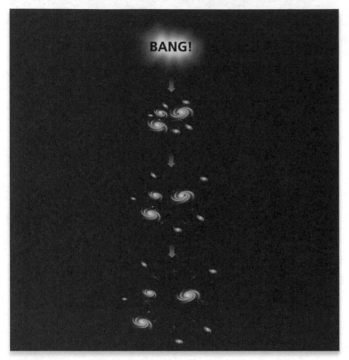

BANG!

The Universe is still expanding after this big explosion. This means that all the galaxies (clusters of stars) are still moving away from us. Scientists have worked out that the furthest galaxies are moving away from us fastest.

We also observe that **microwave radiation** is received from all parts of the Universe.

By tracking the movement of the galaxies, we can estimate the **age** and **starting point** of the Universe.

Evidence for the expansion has also been obtained by the measurement of **red-shift**. If a source of light moves away from us, the wavelengths of the light in its spectrum are longer than if the source were not moving. This effect is known as red-shift because the wavelengths are shifted towards the red end of the spectrum.

Scientists have noted that the light from more distant galaxies is more red-shifted than light from nearer galaxies. This is evidence for the Universe continuing to expand and for galaxies further away moving faster than those closer to us.

By working out how far the light is red-shifted we can work out the different speeds of the galaxies. We can then work backwards to see when the Big Bang happened to give us an estimate of the age of the Universe.

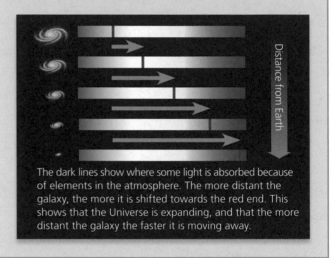

Distance from Earth

The dark lines show where some light is absorbed because of elements in the atmosphere. The more distant the galaxy, the more it is shifted towards the red end. This shows that the Universe is expanding, and that the more distant the galaxy the faster it is moving away.

The History of a Star

Stars, including our Sun, are formed when interstellar (between stars) gas clouds, which contain mainly hydrogen, collapse under gravitational attraction to form a **protostar**. Over a very long period of time, the temperature of the protostar increases as thermonuclear fusion reactions take place, releasing massive amounts of energy, and it finally becomes a **main sequence star** for the majority of its life (like our Sun). During this time, the forces of attraction pulling inwards are balanced by forces acting outwards and the star experiences a normal life. Eventually the supply of hydrogen runs out, causing the death of the star. The type of death depends largely on the star's mass.

The End of a Star

All stars have a finite (limited) life. They start as a huge gas cloud that joins up to form a star – all stars have different sizes. Eventually, the star's supply of fusion fuel runs out and the star swells up, becoming colder and colder to form a **red giant** or **red supergiant**.

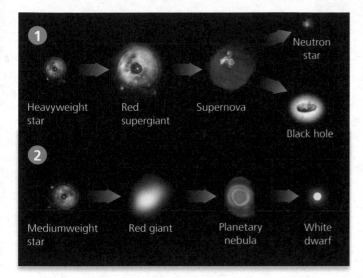

Heavyweight star · Red supergiant · Supernova · Neutron star · Black hole

Mediumweight star · Red giant · Planetary nebula · White dwarf

❶ Death of a Heavyweight Star

Red supergiant to Supernova – when a red supergiant shrinks rapidly and explodes it becomes a supernova, releasing massive amounts of energy, dust and gas into space.

Supernova to Neutron star – for stars up to ten times the mass of our Sun, the remnants of the supernova form a neutron star, which is formed only of neutrons. A cupful of this matter could have a mass greater than 15 000 million tonnes!

Supernova to Black hole – those stars greater than ten times the mass of our Sun are massive enough to leave behind black holes. Black holes can only be observed indirectly through their effects on their surroundings, e.g. the X-rays emitted when gases from a nearby star spiral into a black hole. Not even light can escape from a black hole.

❷ Death of a Mediumweight Star

Red giant to Planetary nebula – For a red giant the core contracts and is surrounded by outer layers (shells) of gas which eventually drift away into space.

Planetary nebula to White dwarf – As the core cools and contracts further, it becomes a white dwarf with a density thousands of times greater than any matter on Earth.

> **HT** Black holes can be found throughout the Universe and in every galaxy. They have a very large mass concentrated into a very small space, which means that their gravity is very large. Not even light can escape from black holes.
>
> Black holes are formed at the end of the life of a heavyweight star.

The History of the Universe

We now know that the Earth orbits the Sun. However, in ancient Greece and ancient China, people believed that the Earth was the centre of the Universe and all the planets orbited around it. All the stars seemed to orbit the Pole Star, while those near the Equator would rise and set. They could not imagine that the Earth was moving.

Nicolas Copernicus in Poland was the first person to publish a book (in 1543) which claimed that the Earth and all the other planets orbited the Sun. This was not accepted because it contradicted the religious view in Europe of how the Christian God made the world.

Galileo Galilei in 1610 used the telescope that he had invented to see that Venus had 'phases' like the Moon. This convinced him that the Earth and other planets orbited the Sun, as Copernicus had said.

> **HT** Copernicus and Galileo contradicted the Roman Catholic Church's view of how the Universe was formed. At this time all the philosophers and astronomers believed in the geocentric view with the Sun orbiting the Earth. The Catholic Church described the heliocentric universe (with the planets orbiting the Sun) as 'false and contrary to scripture'. Galileo was tried and convicted of heresy. This made the heliocentric view very unpopular for many years.

1 This question is about a bicycle dynamo.

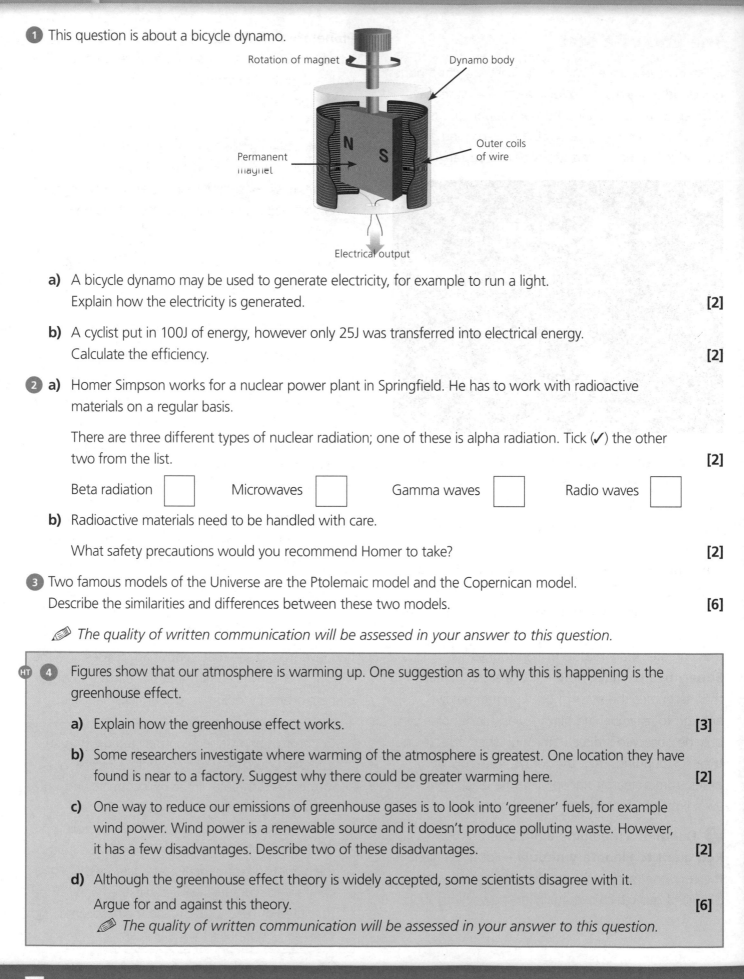

Rotation of magnet

Dynamo body

Permanent magnet

Outer coils of wire

Electrical output

a) A bicycle dynamo may be used to generate electricity, for example to run a light.
Explain how the electricity is generated. **[2]**

b) A cyclist put in 100J of energy, however only 25J was transferred into electrical energy.
Calculate the efficiency. **[2]**

2 **a)** Homer Simpson works for a nuclear power plant in Springfield. He has to work with radioactive materials on a regular basis.

There are three different types of nuclear radiation; one of these is alpha radiation. Tick (✓) the other two from the list. **[2]**

Beta radiation ☐   Microwaves ☐   Gamma waves ☐   Radio waves ☐

b) Radioactive materials need to be handled with care.

What safety precautions would you recommend Homer to take? **[2]**

3 Two famous models of the Universe are the Ptolemaic model and the Copernican model.
Describe the similarities and differences between these two models. **[6]**

✏ *The quality of written communication will be assessed in your answer to this question.*

HT 4 Figures show that our atmosphere is warming up. One suggestion as to why this is happening is the greenhouse effect.

a) Explain how the greenhouse effect works. **[3]**

b) Some researchers investigate where warming of the atmosphere is greatest. One location they have found is near to a factory. Suggest why there could be greater warming here. **[2]**

c) One way to reduce our emissions of greenhouse gases is to look into 'greener' fuels, for example wind power. Wind power is a renewable source and it doesn't produce polluting waste. However, it has a few disadvantages. Describe two of these disadvantages. **[2]**

d) Although the greenhouse effect theory is widely accepted, some scientists disagree with it.

Argue for and against this theory. **[6]**

✏ *The quality of written communication will be assessed in your answer to this question.*

Answers

B1: Understanding Organisms

1. a) A reflex (action)

b) Sensory neurone; Effector/(arm) muscle.

2. a) i) No **[No marks]** Red and crimson are similar colours and Adarsh cannot tell the difference between them.

ii) Yes **[No marks]** Nateem may not inherit the faulty gene at all; She has two X chromosomes; If she does inherit the faulty gene it is unlikely that both chromosomes will carry it/the 'healthy' gene will be expressed **[Any two for 2 marks]**

b) i) Its eyes are on the side of its head.

ii) The rabbit has almost 360° vision and can see behind its head.

3. a) Males have higher percentages of smoking in all years; Rates of smoking are variable between 1948 and 1960; Continual decrease in both sexes since 1960. **[Any two for 2 marks]**

b) Male cancer incidence has have fallen steadily overall; Female cancer incidence has increased since records began.

c) 40 − 25 = 15% **[Accept + or − 2%.]**

d) 1960

4. a) Yes **[No marks]** DDD results in a greater weight loss; 5.8 compared with 3.2; Significantly higher than the placebo **[Any two for 2 marks]**

OR No **[No marks]** Number of volunteers for the DDD trial is very small compared with the other two trials; More trials need to be carried out; Data is unreliable **[Any two for 2 marks]**

b) In a double blind trial neither the volunteers nor the doctors know which pill has been given; This eliminates all bias from the test/scientists cannot influence the volunteers' response in any way.

B2: Understanding our Environment

1. a) Manufacture of protein/DNA/nucleic acid.

b) Photosynthesis

2. a) 87%

b) Yes **[No marks]** – there was a low number of trees with fruticose (and foliose) lichens/high numbers of trees with crustose; **or** No **[No marks]** – there was no comparison of lichen presence in an area without a power station/pine trees had no lichens of any sort.

c) Collect data from an area without a power station; Collect data from more trees/more species of trees; Measure the coverage of trees with lichen – not just their presence. **[Any two for 2 marks]**

3. • **Idea of variation/mutation(s) [1 mark]**
 – allow adapted to environment/had different characteristics
 – allow named examples of different characteristics e.g. some horselike mammals had more flipper-like limbs (as whales have flippers)
• **Idea of competition for limited resources [1 mark]**
 – allow examples of different types of competition
• **Idea of survival of the fittest [1 mark]**
 – allow adaptations were advantageous to living in water/ideas that adaptations helped them survive
 – allow named examples of different adaptations e.g. some horse-like mammals had more flipper like limbs (as whales have flippers) that allowed them to swim well in water
• **Idea of inheritance of successful characteristics [1 mark]**
 – allow (named) characteristics/adaptations passed on (through breeding) **but** 'best adapted breed' is worthy of two possible ideas

4. This is a model answer, which demonstrates QWC and therefore would score the full 6 marks: The polar bear has adapted to live in a habitat of polar ice caps so if the habitat changes,

Answers

the polar bear may not be able to survive. Polar bear numbers would decline. They would no longer be well camouflaged due to there being no snow or ice. Their thick coat and blubber / fat may cause them to overheat in the warmer temperatures and the change to their habitat may lead to a lack of breeding ground. All these factors would lead to a significant overall decline in polar bear numbers.

5. **a)** 120
   b) When kingfishers increase fish decrease/when kingfishers decrease fish increase.
   c) This is a model answer, which demonstrates QWC and therefore would score the full 6 marks: Three rivers is too small a sample so not all birds would be observed, and ringing birds might affect their survival. Some fish may evade capture by anglers and not all anglers will cooperate with the scientists. Not all species will be caught and kingfishers do not feed off every species of fish.

   Improvements could be made by sampling many more rivers, using more observers, sampling fish directly, e.g. mark/recapture technique, by observing which species of fish the kingfishers take, or by using more efficient capture methods for fish e.g. netting.

C1 Carbon Chemistry

1. **a)** Diesel comes from crude oil; This takes millions of years to make.
   b) This is a model answer, which demonstrates QWC and therefore would score the full 6 marks: The terminal would bring investment and jobs to the area improving the living standards of the people who live there. The jobs would be in the building industry and in companies that provide supplies and services to the visiting ships. However, if there is an accident and crude oil spills into the sea then it will float on the water and will damage the environment.

Crude oil is poisonous to wildlife; it will coat the feathers of sea birds stopping them from flying or floating, and may even kill them. The crude oil will float ashore and damage the beaches and shore line.
   c) i) D **ii)** B
   d) Petrol

2. **a)** Non toxic; Easy to control; No pollution; High energy value; Available; No need for storage **[Any two for 2 marks]**
   b) All the fuel is burned in a blue flame (complete combustion) so all the energy is released; Only some of the fuel is burned in a yellow flame so less energy is released.
   c) ethyne + oxygen ➡ carbon dioxide + water

3. Gore-Tex because it is lightweight; Tough; Waterproof; Able to block ultraviolet light; Breathable – allows perspiration to escape whilst preventing rain getting in. **[Any two for 2 marks]**.**(Also accept nylon jacket as long as valid reasons are given, e.g.** It is lightweight; Tough; Waterproof; Able to block ultraviolet light; Cheaper than Gore-Tex jacket **[Any two for 2 marks])**

4. **a)** Ethanol; Ethanoic acid. **[Any one for 1 mark]**
   b) The pigments used.

5. David is correct, except for one value (8km) **[1 mark]** Improvements could be made by: Measuring more than one gas; Monitoring over a longer time period; Calculating or basing claims on average levels. **[Any two for 2 marks]**

6. $CH_4 + 2O_2$ ➡ $CO_2 + 2H_2O$ **[All correct for 2 marks]**

7. There is less photosynthesis; Less carbon dioxide removed from air; Less oxygen made; The burning of the trees puts carbon dioxide into the air; Removes oxygen.

8. The nitrogen is from the air; It combines with oxygen at high temperatures; Use a catalytic converter to remove the nitrogen oxides.

C2 Chemical Resources

1. **a)** The crust and the outer part of the mantle.
   b) It fits a wide range of evidence; It has been tested by many scientists.

2. **a) i)** Noise; Dust; Take up land; Change / destroy the landscape; Increase local traffic. **[Any three for 3 marks]**
      ii) Any suitable answer, e.g. The quarry may provide jobs for local people.
   b) Magnesium carbonate ➔ magnesium oxide + carbon dioxide

3. **a)** It is stronger; It is easier to shape; It is more flexible **[Any two for 2 marks]**
   b) Tin
   c) It is hard; It is strong; It has a low melting point; It is gas proof. **[Any two for 2 marks]**

4. **a)** 550°C
   b) 34%
   c) The yield increases.

5. It contains the essential elements N (nitrogen); and P (phosphorus); It is soluble in water and so can be taken in by plant roots.

6. **a)** Hydrogen; Chlorine; Sodium hydroxide.
   b) $2Cl^- - 2e^- \rightarrow Cl_2$ **[1 mark for correct formulae, 1 mark for correct balancing]**

7. The more dense oceanic plate is pushed under the continental plate; Down into the mantle where it melts; The result is a mountain range and possibly volcanoes.

8. **This is a model answer, which demonstrates QWC and therefore would score the full 6 marks:** The Haber process uses an iron catalyst, which speeds up the rate of reaction so that the ammonia is made as fast as possible. The temperature used is 450°C. This is a compromise because you get a better yield of ammonia if you use a lower temperature, but the ammonia is then made very slowly. At a higher temperature (800°C) the ammonia would be made faster but the yield would be very low. The pressure used is 200 atmospheres because this gives a fast reaction with a good yield

without costing too much to achieve. Using a lower pressure (50atm) would make the reaction slower and reduces the yield.

9. The fertiliser would cause eutrophication; The fertiliser would get into the water and cause algal blooms; This blocks out sunlight; Other plants die; Aerobic bacteria that rot the plant material use up all the oxygen; The water creatures on the reserve would die. **[Any three for 3 marks]**

10. $2HNO_3 + Na_2CO_3 \rightarrow 2NaNO_3 + H_2O + CO_2$
    [1 mark for correct formulae, 1 mark for correct balancing]

P1: Energy for the Home

1. **a)** The ice lolly is melting.
   b) $0.057kg \times 1.34kJ/kg/°C \times 5°C = 0.382kJ$

2. **a)** 30 years; 1.5 years; 20 years
   b) **Accept any suitable suggestion with reasons, e.g.** Double glazing because it has the biggest annual saving; Draught excluders because they have the shortest payback time / lowest cost.
   c) Foil reflects **[1 mark]** heat energy (infrared) **[1 mark]** back into the room / less heat loss so radiators can be turned down. **[1 mark]**

3. **a)** Microwave signals
   b) Speed = frequency × wavelength
   $$Frequency = \frac{3 \times 10^8}{0.02}$$
   $$= 1.5 \times 10^{10}\,Hz$$ **[1 mark for calculation, 1 mark for correct answer]**
   c) The town is next to a hill **[1 mark]** Microwaves are transmitted between transmitters and receivers in line of sight and the hill acts as an obstacle. **[1 mark]**
   d) **Advantage:** Improved mobile phone signal for residents. **[1 mark]**
   Disadvantage: The mast could damage the visual appearance of the area; There are concerns that it could be a danger to health. **[Any one for 1 mark]**

Answers

P2: Living for the Future (Energy Resources)

1. **a)** The coil rotates/moves near to the magnet **[1 mark]** which induces (generates) an electric current in the wire. **[1 mark]**

    b) $\dfrac{\text{Energy out}}{\text{Energy in}} \times 100\%$

    $\left(\dfrac{25}{100}\right) \times 100\% = 25\%$ **[1 mark for calculation, 1 mark for correct answer]**

2. **a)** Beta radiation **and** Gamma waves. **[1 mark each]**.

    b) Wear protective clothing; Use tongs/keep your distance; Short exposure time; Shielded and labelled storage. **[Any two for 2 marks]**.

3. **This is a model answer, which demonstrates QWC and therefore would score the full 6 marks:** The Ptolemaic and Copernican models are similar in that they both proposed that the planets were represented by glass spheres a fixed distance from the Sun, and that the stars were in fixed positions on the outermost sphere.
    The two models are different in that the Copernican model stated that the Sun was at the centre of the Universe, the Earth rotates once every 24 hours and the Earth takes one full year to revolve around the Sun. In contrast, the Ptolemaic model stated that the Earth was the centre of the Universe.

4. **a)** Short wavelength electromagnetic radiation from the Sun is absorbed by and heats the Earth **[1 mark]**; The Earth radiates the heat as longer wavelength infrared radiation **[1 mark]**; Greenhouse gases absorb some infrared radiation, which is re-emitted back to Earth, warming the atmosphere. **[1 mark]**

    b) Dust **[1 mark]** released into the atmosphere from the factory would cause radiation from a nearby town to be reflected back to the Earth, causing warming. **[1 mark]**

    c) Visual pollution; Dependent on wind speed; Appropriate space and position needed. **[Any two for 2 marks]**

    d) **This is a model answer, which demonstrates QWC and therefore would score the full 6 marks:** There has been a measured increase in the global temperature and an increase in the level of greenhouse gases such as carbon dioxide and methane in the atmosphere, which leads some to conclude that the greenhouse gases are causing the global temperature increase. Others indicate that global temperatures have increased (and decreased) on many occasions in the past and that there is evidence that solar activity is responsible. There is no doubt that global temperatures are rising but the disagreement is whether greenhouse gases are the cause.

Glossary

Abundance – the number of organisms in a habitat.

Acid – a substance that dissolves in water to give a solution with a pH value lower than 7.

Addiction – a craving for a chemical/drug, either psychological or physical.

Alkali – a substance that dissolves in water to give a solution with a pH value higher than 7.

Alkane – a hydrocarbon molecule containing single bonds only.

Alkene – a hydrocarbon molecule containing a double bond.

Allele – one of a pair of alternative genes on homologous chromosomes.

Alloy – a mixture of two or more metals, or of a metal and a non-metal.

Alpha radiation – nuclear radiation particle made up of 2 protons and 2 neutrons.

Alternating current (AC) – an electric current that changes direction of flow repeatedly.

Amino acid – building block of protein molecules.

Amplitude – the maximum disturbance of a wave from a central position.

Analogue – signal that varies continuously in amplitude or frequency.

Anode – the positive electrode.

Antibiotic – anti-bacterial drug.

Antibody – proteins produced by the immune system that neutralise pathogens.

Antigen – marker molecule on surface membrane of cell. Recognition point for the immune system, e.g. on a pathogenic microbe.

Asteroid – large rock (smaller than a planet) that orbits the Sun in a belt between Mars and Jupiter – occasionally knocked off course by other rocks.

Artery – large blood vessel with narrow lumen and thick elastic walls.

Atom – the smallest part of an element that can enter into chemical reactions.

Base – a substance that will neutralise an acid.

Beta radiation – nuclear radiation made up of a fast moving electron.

Big Bang theory – theory of how the Universe started.

Biomass – natural materials such as wood or manure that can be burned or fermented to produce methane as a fuel.

Black hole – formed at the end of a star's life; has very dense core which exerts extreme gravity so that even light cannot escape.

Capillary – very small blood vessels forming a network around all body tissues.

Carbohydrate – energy-rich nutrient found in sugary and starchy foods.

Catalyst – a substance that is used to speed up a chemical reaction without being chemically changed at the end of the reaction.

Cathode – the negative electrode.

Chromosome – structure found in the nucleus of the cell carrying genetic information.

Climate change – changes in climate that result in increasing average global temperatures.

Comet – orbits the sun in an elliptical orbit; has frozen gas and dust core.

Compound – a substance consisting of two or more atoms chemically combined together by ionic or covalent bonds.

Conduction – transfer of thermal or electrical energy.

Conductor – material that transfers thermal or electrical energy.

Consumer – an organism in a food chain that eats another organism.

Covalent bond – a bond between two atoms formed by sharing a pair of electrons.

Cracking – a process used to break up large hydrocarbon molecules into smaller, more useful molecules.

Critical angle – the largest angle of incidence within a medium at which refraction can occur.

Current – the rate of flow of an electrical charge; measured in amperes (A).

DAB (digital audio broadcasting) – transmitting radio signals using digital signals rather than analogue signals.

Decomposer – organism that breaks down dead organic matter.

Decomposition – the breaking down of a substance into simpler substances, e.g. using heat or electricity.

Deforestation – removal of large areas of forest.

Degrees Celsius (°C) – unit of temperature.

Diabetes – condition caused by inability to produce enough of the hormone insulin.

Digital – signal that uses binary code (ons and offs such as Morse code).

Direct current (DC) – an electric current that flows only in one direction.

Distribution – the spread of organisms in a habitat.

DNA – large molecule carrying a genetic code that makes up chromosomes.

Dynamo effect – generating electricity by moving a coil of wire near a magnet.

Effector – organ/tissue that carries out the response of a nervous pathway, usually a muscle.

Efficiency – useful output energy expressed as a percentage of total input energy.

Electric current – flow of charge through conductors in a circuit.

Electrode – the conducting rod or plate (usually metal or graphite) that allows electric current to enter and leave an electrolysis cell.

Electrolysis – the breaking down of a liquid ionic substance using electricity.

Electrolyte – an aqueous or molten substance that contains free-moving ions and is therefore able to conduct electricity.

Electromagnetic waves / spectrum – includes radio waves, visible light and gamma radiation, all of which travel through a vacuum at the speed of light.

Electron – a negatively charged particle that orbits the nucleus of an atom; a charged particle that flows through wires as an electric current.

Element – a substance that consists of only one type of atom.

Emulsion – a mixture of one liquid finely dispersed in another liquid.

Endocrine gland – ductless gland that produces hormones

Energy – measure of ability to do work or of heat transferred; measured in joules (J).

Enzymes – molecules acting as biological catalysts.

Evolution – slow progress of change in organisms' structure over millions of years.

Fossil fuel – a substance that is burned to release heat energy; formed from the remains of plants and animals over millions of years.

Fraction – a mixture of hydrocarbons with similar boiling temperatures that separated during distillation.

115

Glossary

Frequency – of AC – the number of cycles completed each second; of waves – the number of waves produced (or that pass a particular point) in one second.

Gamete – sex cell; sperm or egg/ovum.

Gamma radiation – nuclear radiation that is high frequency electromagnetic waves.

Geiger counter – a device for measuring radioactivity.

Gene – short section of a chromosome that codes for a protein molecule.

Generator – a device for making electric current using a magnet and a coil of wire.

Global warming – raising of average global temperatures as a result of the greenhouse effect.

Glucose – basic sugar molecule used for release of energy in organisms.

Gravitational potential energy – the energy an object has because of its mass and height above Earth.

Gravity (gravitational force) – a force of attraction between masses.

Greenhouse effect – trapping of heat by the atmosphere.

Group – a vertical column of elements in the Periodic Table.

Haber process – process used to make ammonia.

Homeostasis – maintenance of a constant internal environment.

Hydrocarbon – a molecule containing hydrogen and carbon only.

Hydrophilic – water loving.

Hydrophobic – water hating.

Immunity – state of the human body when it is protected from infection by action of white blood cells and antibodies.

Incident angle – angle measured from the normal at which light approaches a boundary between two materials.

Indicator species – organisms acting as 'marker' species for detecting environmental pollution.

Infrared – hotter objects emit more infrared radiation than cooler objects. Infrared is a form of electromagnetic radiation.

Inheritance – the passing of characteristics from parents to offspring in the form of genes.

Insulator – a substance that does not transfer thermal or electrical energy very well.

Interference – on a radio the hissing noise is interference as a result of the signal being corrupted; two waves can interfere and either reinforce each other or cancel out.

Ion – a positively or negatively charged particle formed when an atom gains or loses one or more electron(s).

Ionic bond – the bond formed when electrons are transferred between a metal atom and a non-metal atom, creating charged ions that are held together by forces of attraction.

Ionising – radiation that turns atoms into ions.

Joule (J) – unit of energy.

Kilowatt hour – a measure of how much electrical energy has been used.

Kinetic energy (KE) – the energy possessed by a body because of its movement.

Lichen – plant-like structure resulting from a mutualistic relationship between algae and fungi.

Longitudinal wave – an energy-carrying wave where the particles of the medium move in the direction of energy transfer.

Magnetic field – the area of effect of a magnet (or the Earth) indicated by lines of force surrounding the magnet (or the Earth).

Melt – to change from solid to liquid.

Meteor – small rock orbiting a star.

Microwaves – electromagnetic waves that can be used for transmitting messages and for cooking.

Molecule – two or more atoms bonded together.

Neurone – a nerve cell.

Neutron – a particle found in the nucleus of atoms; it has no charge; relative mass 1.

Neutron star – star made of neutrons.

Non-renewable – made at a slower rate than it is used up.

Nuclear power – generating electricity using uranium or plutonium as the fuel to transfer heat.

Nuclear radiation – alpha or beta particles or gamma radiation given off from radioactive materials.

Nucleus – the core of an atom, made up of protons and neutrons (except hydrogen, which contains a single proton).

Optical fibre – thin glass fibre that carries digital signals.

Orbit – the path of an object around a larger object.

Orbital period – the time it takes an object to make one complete orbit.

Ozone – a gas in the upper atmosphere.

Oxidation – a reaction involving the gain of oxygen or the loss of electrons.

Pathogen – organism that causes harm or death; usually disease-causing microbes.

Payback time – the time taken for insulation to pay for itself from savings made.

Period – a horizontal row of elements in the Periodic Table.

Periodic Table – a list of the elements arranged to show the trends and patterns in their properties.

Photocell – a device that captures light energy and transforms it into electrical energy.

Plutonium – radioactive fuel used in some nuclear power stations.

Polymerisation – the reaction of many monomer molecules joining together to make one large polymer molecule.

Population – number of individuals of a single species found in a given area.

Power – the rate of doing work; measured in watts (W).

Primary waves (P-waves) – earthquake waves that travel through the Earth arriving at the detector first.

Producer – first organism in a food chain; manufactures its own food, usually a plant.

Product – a substance produced in a reaction.

Progesterone – female hormone produced from the ovary; maintains the uterus lining.

Proton – a positively charged particle found in the nucleus of an atom; relative mass 1.

Protozoan – single-celled microorganism; more complex than bacteria or viruses.

Pyramid of biomass – a diagram showing the mass of organisms in a food chain.

Pyramid of numbers – a diagram showing the numbers of organisms in a food chain.

Radiation – electromagnetic waves or particles emitted by a radioactive substance; transfer of heat as infrared electromagnetic radiation.

Reactant – a starting material in a reaction.

Glossary

Receptor – organ or tissue that detects stimuli, e.g. eye.

Red giant – in their life cycle some stars will explode to form a red giant.

Reduction – a reaction involving the loss of oxygen or the gain of electrons.

Reflection – change in direction of a wave back into a medium at a boundary between two media.

Refraction – change in direction of a wave as it passes from one medium to another and changes speed.

Resistance – how hard it is to get a current through a component at a particular potential difference; measured in ohms (Ω).

Salt – the product of a chemical reaction between a base and an acid.

Satellite – an object that orbits a planet, e.g. the Moon

Saturated – a compound in which all carbon–carbon bonds are single bonds.

Secondary waves (S-waves) – earthquake waves that travel through the Earth and arrive second at the detector.

Seismic wave – wave produced by an earthquake.

Seismometer – machine used to detect seismic waves.

Soluble – a property that means a substance can dissolve in a solvent.

Solute – a substance that gets dissolved by a solvent.

Solution – the mixture formed when a solute dissolves in a solvent.

Solvent – a liquid that can dissolve another substance to produce a solution.

Species – smallest taxonomic group or type; second part of binomial name.

Specific heat capacity – value of how much energy a material requires, per kilogram, to raise its temperature by 1°C.

Specific latent heat – heat energy required to melt or boil 1kg of a material.

State (of matter) – whether a material is solid, liquid or gas.

Supernova – an exploding star.

Synthetic – made or manufactured; not natural.

Temperature – the degree of hotness of an object measured in degrees Celsius (°C).

Thermogram – image showing the different temperatures of an object through different colours.

Tolerance – state in the body in which a greater dose of a drug is required to produce the same effect.

Toxins – poisonous substances produced by pathogens.

Total internal reflection – complete reflection of a light or infrared ray back into a medium.

Transfer – move energy from one place to another.

Transform – change energy from one form to another, e.g. kinetic to electrical.

Transverse wave – a wave in which the vibrations are at 90° to the direction of energy transfer.

Tropism – growth response of plants, e.g. phototropism – light response.

Universal indicator – a mixture of pH indicators, which produces a range of colours according to pH and can therefore be used to measure the pH of a solution.

Uranium – radioactive material used as fuel in a nuclear power station.

Unsaturated – a compound in which at least one carbon–carbon bond is a double bond.

Vector – organism (usually bacterium) which carries DNA for transferring to another organism. Alternatively it can be an insect that carries a parasite and transfers it to another organism.

Vein – large blood vessel with relatively thin walls and valves.

Vitamin – micronutrient needed in the diet for general health.

Voltage (potential difference) – the difference in potential between two points in an electrical circuit; the energy transferred in a circuit by each coulomb of charge; measured in volts (V).

Watt – unit of power.

Wavelength – the distance between corresponding points on two adjacent disturbances.

White dwarf – in the life cycle of a star, a medium-sized star will collapse to form a white dwarf.

Yield – the amount of product obtained, e.g. from a crop or a chemical reaction.

HT

Adipose tissue – tissue storing lipid/fat material, particularly in skin.

Carbon sink – part of the environment that absorbs carbon and locks it in.

Centripetal force – the external force towards the centre of a circle required to make an object follow a circular path at a constant speed.

Convection – transfer of heat energy involving the movement of the substance.

Diffraction – the spreading out of a wave as a result of passing an obstacle or passing through a gap.

Generalist – an organism with a wide range of adaptations.

Genotype – the combination of genes which code for a characteristic.

Heterozygous – pair of alleles on homologous chromosomes that are different.

Homozygous – pair of alleles on homologous chromosomes that are identical.

Intermolecular – describes the force of attraction between one molecule and another.

Ionosphere – band of charged particles (ions) in the Earth's atmosphere.

Light year – measure of distance; travelled by light in a year.

Monochromatic – monochromatic light has a single wavelength.

Multiplex – sending more than one signal along an optical fibre at a time, using different wavelengths.

Niche – small part of the ecosystem that an organism occupies in which it carries out its role.

Phase – two waves are in phase if they have peaks and troughs occurring together.

Phenotype – the appearance of a characteristic in an organism as a result of genetics and environment.

Specialist – an organism with a narrow range of adaptations.

Subduction – a plate boundary where one tectonic plate is forced below the other and the rock melts into the magma.

Synapse – junction between two or more nerve cells.

Vasoconstriction – narrowing of small arteries.

Vasodilation – widening of small arteries.

Volatile – describes a substance that evaporates quickly.

Periodic Table

Key

relative atomic mass
atomic symbol
name
atomic (proton) number

				1	H hydrogen 1		

1	2												3	4	5	6	7	0
																		4 **He** helium 2
7 **Li** lithium 3	9 **Be** beryllium 4												11 **B** boron 5	12 **C** carbon 6	14 **N** nitrogen 7	16 **O** oxygen 8	19 **F** fluorine 9	20 **Ne** neon 10
23 **Na** sodium 11	24 **Mg** magnesium 12												27 **Al** aluminium 13	28 **Si** silicon 14	31 **P** phosphorus 15	32 **S** sulfur 16	35.5 **Cl** chlorine 17	40 **Ar** argon 18
39 **K** potassium 19	40 **Ca** calcium 20	45 **Sc** scandium 21	48 **Ti** titanium 22	51 **V** vanadium 23	52 **Cr** chromium 24	55 **Mn** manganese 25	56 **Fe** iron 26	59 **Co** cobalt 27	59 **Ni** nickel 28	63.5 **Cu** copper 29	65 **Zn** zinc 30		70 **Ga** gallium 31	73 **Ge** germanium 32	75 **As** arsenic 33	79 **Se** selenium 34	80 **Br** bromine 35	84 **Kr** krypton 36
85 **Rb** rubidium 37	88 **Sr** strontium 38	89 **Y** yttrium 39	91 **Zr** zirconium 40	93 **Nb** niobium 41	96 **Mo** molybdenum 42	[98] **Tc** technetium 43	101 **Ru** ruthenium 44	103 **Rh** rhodium 45	106 **Pd** palladium 46	108 **Ag** silver 47	112 **Cd** cadmium 48		115 **In** indium 49	119 **Sn** tin 50	122 **Sb** antimony 51	128 **Te** tellurium 52	127 **I** iodine 53	131 **Xe** xenon 54
133 **Cs** caesium 55	137 **Ba** barium 56	139 **La*** lanthanum 57	178 **Hf** hafnium 72	181 **Ta** tantalum 73	184 **W** tungsten 74	186 **Re** rhenium 75	190 **Os** osmium 76	192 **Ir** iridium 77	195 **Pt** platinum 78	197 **Au** gold 79	201 **Hg** mercury 80		204 **Tl** thallium 81	207 **Pb** lead 82	209 **Bi** bismuth 83	[209] **Po** polonium 84	[210] **At** astatine 85	[222] **Rn** radon 86
[223] **Fr** francium 87	[226] **Ra** radium 88	[227] **Ac*** actinium 89	[261] **Rf** rutherfordium 104	[262] **Db** dubnium 105	[266] **Sg** seaborgium 106	[264] **Bh** bohrium 107	[277] **Hs** hassium 108	[268] **Mt** meitnerium 109	[271] **Ds** darmstadtium 110	[272] **Rg** roentgenium 111								

Elements with atomic numbers 112–116 have been reported but not fully authenticated

*The lanthanoids (atomic numbers 58–71) and the actinoids (atomic numbers 90–103) have been omitted.